T0238699

Lecture Notes in Computer Science   11390

Commenced Publication in 1973
Founding and Former Series Editors:
Gerhard Goos, Juris Hartmanis, and Jan van Leeuwen

More information about this series at http://www.springer.com/series/8637

Abdelkader Hameurlain
Roland Wagner · Tran Khanh Dang (Eds.)

Transactions on Large-Scale Data- and Knowledge- Centered Systems XLI

Special Issue on Data and Security Engineering

Springer

Editors-in-Chief
Abdelkader Hameurlain
IRIT, Paul Sabatier University
Toulouse, France

Roland Wagner
FAW, University of Linz
Linz, Austria

Guest Editor
Tran Khanh Dang
Ho Chi Minh City University of Technology
Ho Chi Minh City, Vietnam

ISSN 0302-9743 ISSN 1611-3349 (electronic)
Lecture Notes in Computer Science
ISSN 1869-1994 ISSN 2510-4942 (electronic)
Transactions on Large-Scale Data- and Knowledge-Centered Systems
ISBN 978-3-662-58807-9 ISBN 978-3-662-58808-6 (eBook)
https://doi.org/10.1007/978-3-662-58808-6

Library of Congress Control Number: 2019931530

This Springer imprint is published by the registered company Springer-Verlag GmbH, DE
part of Springer Nature
The registered company address is: Heidelberger Platz 3, 14197 Berlin, Germany

Preface

The 4th International Conference on Future Data and Security Engineering (FDSE) took place from November 29 to December 1, 2017, in Ho Chi Minh City, Vietnam, at the HCMC University of Technology, among the most famous and prestigious universities in Vietnam. The annual FDSE conference is a premier forum designed for researchers, scientists, and practitioners interested in state-of-the-art and state-of-the-practice activities in data, information, knowledge, and security engineering to explore cutting-edge ideas, to present and exchange their research results and advanced data-intensive applications, as well as to discuss emerging issues on data, information, knowledge, and security engineering. At the annual FDSE, researchers and practitioners are able to not only share research solutions to problems of today's data and security engineering themes, but also to identify new issues and directions for future related research and development work.

We encouraged the submission of both original research contributions and industry papers. The call for papers resulted in the submission of 128 papers. A rigorous peer-review process was applied to all of them. This resulted in 28 full (including keynote speeches) and seven short accepted papers (acceptance rate: 27.3%), which were presented at the conference. Every paper was reviewed by at least three members of the international Program Committee, who were carefully chosen based on their knowledge and competence. Among the great papers of FDSE 2017, we selected nine papers to invite the authors to revise, extend, and resubmit for publication in this special issue. Finally, only seven extended papers were accepted. The main focus of this special issue is on data and security engineering, as well as emerging applications.

The big success of FDSE 2017 and this special issue of TLDKS was the result of the efforts of many people, to whom we would like to express our gratitude. First, we would like to thank all the authors who extended and submitted papers to this special issue. We would also like to thank the members of the committees and external reviewers for their timely reviewing and lively participation in the subsequent discussion in order to select the high-quality papers published in this issue. Finally, yet importantly, we thank Gabriela Wagner for her enthusiastic help and support during the preparation process for this publication.

November 2018   Tran Khanh Dang

Organization

Editorial Board

Trong Nhan Phan   HCMC University of Technology, VNUHCM,
                             Vietnam
Michel Toulouse   Vietnamese-German University, Vietnam/Germany
Minh Quang Tran   HCMC University of Technology, VNUHCM,
                             Vietnam
Anh Truong   HCMC University of Technology, VNUHCM,
                             Vietnam, and Trento University, Italy
Quang Hai Truong   Singapore Management University, Singapore

Contents

Fast Distributed Top-q and Top-k Query Processing

Claus Dabringer and Johann Eder$^{(\boxtimes)}$

Department of Informatics-Systems,
Alpen-Adria Universität, Klagenfurt, Austria
{claus.dabringer,johann.eder}@aau.at

Abstract. Top-k queries retrieve the k results of a query which score best for an objective function representing the preferences of users. To require that the returned results also have to satisfy the preferences to a certain degree we introduce top-q queries which return all results which approximate the user preferences to at least some minim degree q. We show how top-q queries and top-k queries can be combined enabling the user to post a large number of interesting queries. Furthermore, we show that the calculation of top-q queries can be integrated in algorithms efficiently processing top-k queries. We implemented our approach and evaluated it against the fastest threshold based top-k query answering approaches (BPA-2). Our experiments showed an improvement by one to two orders of magnitude regarding time and memory requirements. Furthermore, we show how such queries can be processed in highly distributed peer-to-peer databases in an efficient way and propose an adaptive algorithm which takes several parameters of the network of databases into account to optimize the processing of distributed top-k queries.

Keywords: Top-q query answering · Top-k query answering ·
Approximate querying · Result ranking · Distributed Top-k queries ·
Adaptive query processing · p2p databases

1 Introduction

Users searching in large and multi-faceted databases like product catalogs, hotel databases, real-estate databases, or scientific database are often confronted with the problem, that classical queries do not distinguish between hard and soft restrictions. Soft restrictions express the preferences of the users while hard restrictions express necessary constraints. In traditional query evaluation such queries with all preferences stated as hard constraints return few or no results, probably missing acceptable answers when users do not expect that all their - possibly conflicting - preferences are fulfilled. If all the soft restrictions are relaxed, the answer set is possibly huge. Therefore, to achieve satisfying results,

The work reported here was supported by the Austrian Ministry for Science and Research within the projects GATIB II and BBMRI.AT.

© Springer-Verlag GmbH Germany, part of Springer Nature 2019
A. Hameurlain et al. (Eds): TLDKS XLI, LNCS 11390, pp. 1–31, 2019.
https://doi.org/10.1007/978-3-662-58808-6_1

users have to repeat the queries iteratively slightly relaxing some of the restrictions each time. The main application area in our focus is searching biobanks [5,8,15,17,18,27,27,32,41] for samples which are suitable for an intended medical study.Frequently, the ideal sample is overspecified and we observed, how users manually checked very long result lists or repeated queries many times slightly varying several restrictions.

Top-k queries provide an approach to efficiently express such queries: they return the k tuples, which score best for an objective function. But plain top-k query answering also has its limitations: some, many or even all of the returned objects may not be similar enough to the ideal object to be useful for the user. This leads to unnecessary efforts, in particular, when top-k queries are executed over distributed databases [13]. To overcome this problem we propose top-q queries which retrieve all objects within a certain similarity range q around the objects specified through a query. We show how top-q and top-k queries can be combined, e.g., to retrieve the best 100 objects that are at least 80% similar to an ideal object.

While it is conceptually trivial to formulate such queries in SQL, classical relational query processing would be very inefficient, wasting enormous resources. The efficient retrieval of the top-k tuples approximating an ideal object has, therefore, attracted a lot of research attention (see [30] for an overview). We present an algorithm using elaborate indexing techniques [12], which can processes both top-k and top-q queries as well as combinations of top-k and top-q.

Optimizing top-k queries in highly distributed networks of federated or peer-to-peer (p2p) databases still has significant research needs to consider both query response time and system effort.

Processing a top-k query in a p2p network with horizontal partitioning involves sending a top-k_p query to each peer p. The optimization problem now is to determine a proper k_p for each peer p, i.e. how many objects should be fetched from which peer. If k_p is too large, it results in unnecessary computation at the peers' sites and unnecessary traffic. If it is too low, it is necessary to send additional queries to the peers. We propose an adaptive approach considering both objectives.

According to the categories introduced by Ilya et al. [30] our algorithm can be classified as follows: *Query Model*: top-q selection and top-q join, *Data & query certainty*: certain data, exact methods, *Data access*: both sorted and random, *Implementation level*: application level, *Ranking function*: monotone.

We reported on these approaches and results already in conferences. This paper is an extension of [13] and [14] integrating the top-q, top-k, and distributed top-k approaches.

The rest of the paper is organized as follows: In Sect. 2 we discuss the principles of processing top-k queries, related work, and areas for improvements. We present our *TQQA* approach for top-q and top-k queries in Sect. 3. In Sect. 4 we present our *ADiT* approach for efficiently processing top-k queries in a distributed network of databases. Implementations of the *TQQA* and the *ADiT* approach together with extensive experimental evaluations are given in the Sects. 5 and 6. In Sect. 7 we draw some conclusions.

2 Top-k Query Answering

The well known and widely applied Threshold Algorithm of Fagin et al. (in short TA) [19] (independently published by Guntzer et. al [24] and Nepal et al. [35]) uses a *pre-calculated index structure* for evaluation of similarity. It basically assumes m grades for each object in a database where m is the number of different attributes. Fagin assumes for every attribute a sorted list which stores each object and its similarity under a certain attribute value. An example list is shown in Table 1. TA processes the given lists row by row and maintains a buffer of objects most similar to the ideal object specified by the input query. Since the lists are ordered the algorithm stops in case k-objects have been identified and no unseen object has the chance of achieving a higher score than the object with the lowest score stored in the buffer. There exist several different versions and adaptions (*no random access - NRA, Stream-Combine approach, LARA*) [19,25,33] of the original TA algorithm which work on different assumptions but do not improve query answering speed considerably. On the other hand the *best position algorithm* BPA-2 [4] relies on the same assumptions as TA does, but incorporates an earlier stopping condition. For each sorted list BPA-2 maintains the so called *best position*, i.e. the greatest position such that any lower position in the list has already been examined. In each step BPA-2 examines the next position after the best position in each list and does random access to all other lists to find the similarity values of the processed object there. With that technique BPA-2 achieves two improvements over TA: (1) it avoids duplicate access to one and the same object and (2) it can stop much earlier than TA does. [4] showed that BPA-2 can improve query answering performance of TA up to eight times.

Table 1. Sorted list for Attr-1

Object ID	Attr-1 grade
5	50
1	40
3	30
2	20
4	10
...	...

A good overview on top-k queries is given in [30]. The efficient answering of queries searching for the top-k most typical tuples in large databases is discussed in [29]. The authors introduce new query operators to enhance the existing SQL syntax. The works [4,33] can be classified in the same category as Fagin's TA. They all rely on *m-sorted lists* which contain each object stored in a certain database with its corresponding rank. Since some query answering systems integrate their objective functions into the core of the database it is often needed to rewrite that functions when needs change. In [31] a generic system is presented which allows to define objective functions outside the database core, thus there are no needs to rewrite existing code when requirements change. Ranking of queries in systems supporting only boolean queries is covered in [28].

Our approach is inspired by the idea of Fagin's TA [19] to maintain a set of indexes and process them. In [11] the authors already showed that the usage of appropriate indexing techniques can improve query answering times and memory requirements by a factor of two to five. Opposed to object-based indexes the authors recommended to generate an index entry for each distinct attribute value and proposed a parallel stepwise processing of the generated indexes. In [12] it is shown how an intelligent processing of the indexes can lead to an additional significant reduction in query answer time as well as in memory requirements. The parallel stepwise processing (as done in TA and BPA-2) of the index lists leads to examining many objects which do not make it in the answer set. Therefore, it was suggested to apply an intelligent look-ahead technique which always chooses to process the most promising index structure next.

Distributed top-k query processing did not receive as much consideration as centralized top-k query processing. There are approaches which optimize a particular measure, respectively take particular configuration parameters into account. There are approaches which focus on reducing the amount of transmitted queries [3,26]. Other approaches aim at keeping the amount of transported objects low [6,9,21,38]. Yet other approaches strive to reduce the communication costs [20]. However, both the transmitted objects and messages affect the *system effort* and *query response time* in a peer-to-peer system.

3 Top-Q Query Answering

The *TQQA* approach is able to find all objects that do not fall below a certain percentage threshold compared to the ideal object expressed through the input query.

3.1 Query Formulation

TQQA supports a great range of queries which may contain an arbitrary number of attributes from different tables as well as restrictions affecting attributes from different tables. Generically, the supported queries can be written as follows:

$$
\begin{aligned}
&select\ a_1,\ a_2,\ ...\ a_n \\
&\quad from\ t_1,\ t_2,\ ...\ t_t \\
&\quad where\ join_1\ ...\ AND\ join_j \\
&\qquad AND\ restriction_1 \\
&\qquad AND\ restriction_2\ ... \\
&\qquad AND\ restriction_r
\end{aligned}
$$

The restrictions can be any binary operator in standard SQL. Restrictions composed of binary operators consist of a left-hand attribute, an operator and a right-hand attribute or value, e.g. *age = 30*. Additionally, we introduce the operator '~', indicating soft restrictions.

It is also possible to combine top-q queries with existing top-k queries. One can define a threshold value q and specify to retrieve *at least/at most* k-objects leading to four possible scenarios for retrieving the top $k = 100$ objects for a percentage threshold of q = 90%, i.e. retrieving of:

- *at most* 100 objects → returns 100 objects
- *at most* 1000 objects → returns 500 objects
- *at least* 100 objects → returns 500 objects
- *at least* 1000 objects → returns 1000 objects.

Necessary Functions. One of the basic concepts of *TQQA* is its flexibility since users may customize and thus affect the ranking by providing a user defined *similarity and an objective function*. *TQQA* basically requires these two functions which can be defined by the user (for convenience we provide useful default functions):

1. Similarity Function: The similarity function is used to calculate an index structure in the initialization phase. It must be monotonous and able to calculate the similarity between two values of an attribute in a database or in a relation. To support a broad field of applications it is desirable to provide different similarity functions for different attribute types, e.g. numerical values, categorical values or string values. It is also possible to bind a similarity function to a column of a relation or view. This allows the user to have different similarity functions for different attributes, e.g. age and weight might have different functions to calculate the similarity of their values.
2. Objective Function: The objective function is used for calculating the rank or score of objects. It defines how well an object satisfies the given restrictions, i.e. how similar an object's attribute values are to the values in the restrictions. With this function TQQA calculates the maximum possible score a (possible hypothetical) object in the database can achieve. Then we only return objects that are at least q-% similar to that maximum possible score. *TQQA* also supports the definition of objective functions with more than one input parameter, e.g. it is possible to pass weights etc. into the objective function. TQQA (as all other threshold algorithms) requires monotonous objective functions.

These two functions can be passed to TQQA in an initialization step which precedes the actual searching for interesting objects. Both functions are *not restricted* to return values in a specific interval only, e.g. 0 to 100. An implicit scaling to a percentage value is provided by the TQQA algorithm. Within this section we provided the description of the two functions from a meta-level. A concrete example for each function is given in Sect. 5 where we instantiate the *TQQA* approach.

3.2   Processing TQQA Queries

The goal of developing algorithms for processing $TQQA$ queries is to derive the correct result without retrieving all the objects from the database. In this section we show how an efficient algorithm developed for the processing of top-k queries can be adopted to process top-q queries and combined top-k and top-q queries in an efficient way. The presented technique can also be applied to accelerate other types of queries with restrictions on monotone functions [16].

In contrast to traditional top-k approaches [4,19,24,34] which maintain object indexes TQQA uses a sorted similarity list (i.e., an index) for any distinct attribute value computed with the given similarity function. Since the index contains distinct attribute values only, the memory requirements of the TQQA approach are proportional to the selectivity of the restricted attributes. This leads to a significant speedup compared to approaches which calculate similarity measures for each object under each attribute [11]. In addition, we make use of the look-ahead technique presented in [12], which always processes the most promising index next.

The $TQQA$ approach operates in two phases, (1) an *initialization phase* and (2) a *processing phase*. Within the *first phase* TQQA is initialized with two functions, namely the *similarity function* and the *objective function* and creates an *index lookup structure* based on the given restrictions and calculates the maximum possible score. The *second phase* processes the index lists and selects objects with minimum similarity. It stops when no more objects with that minimum similarity can be obtained from the database or when the enough objects are published. The following piece of pseudo-code shows the process of obtaining the top-q objects for a given list of restrictions.

```
program Top-QK (IN string tableName, IN Set restr,
                IN Number q, IN Number K,
                IN StoppingType type
                IN Function Sim, IN Function Obj,
                OUT Map<Number, object> objects)
    var idx: Set<LookupTable>
    var maxScore: Number
    var buf: Map<Number, object>
begin
    SetSimilarityFunction(Sim);
    SetObjFunction(Obj);
    CreateIndexTables(restr, OUT idx);
    GetObjectScore(to_object(restr), idx,
                    OUT maxScore);
    while contSearch(idx, maxScore * q, K, type)
      processNext(tableName, idx, I_O buf);
      publishObjects (buf, maxScore * q,
                      K, type, I_O objects)
    end-while
end.
```

Seven input parameters are needed: (1) The *tableName* is the name of the table containing the data to be searched. If more than one table is involved a view joining all needed tables is generated. (2) The restrictions which should be satisfied by the result tuples. (3) The parameter q defining minimum similarity. (4) The parameter k specifying the number of objects. (5) The *StoppingType*(*at least* or *at most*) specifying the combination of q and k. (6) The reference to the similarity function. (7) A reference to the objective function. The output produced is a sorted list of objects.

In the following we take a deeper look into the major parts of the *TQQA* approach. For examples, we assume a relation R which contains attributes *a1 (number of seats)*, *a2 (type of car)*.

Generation of Index Lookup Tables. The final step in the initialization phase is the generation of a set of index lookup tables. TQQA generates an index for each restriction in the input set directly through an SQL command as shown in pseudo-code:

```
program CreateIndexTables (IN Set restrictions,
                           OUT Set<LookupTable> idx)
    var q: Query
    var lookup: LookupTable
begin
    foreach restriction r in restrictions
      q := select distinct(r.attr),
             sim(r.val, r.attr) similarity
             from r.relation
             order by similarity desc;
      execute immediate q bulk collect into lookup;
      idx.Add(lookup);
    end-for
end.
```

Below we show how this algorithm works in our example ex-1. We assume the user specified the following restrictions: "a1 ~5 AND a2 ~sedan".

When filling in the values of the example restrictions into the query this leads to the following query statement ExQ-1:

> *select distinct(a1),*
>   *sim (5, a1) similarity*
>   *from R order by similarity desc;*

This query retrieves a similarity list of all distinct values from attribute *a1* from relation R. The reference value for calculating the similarity is 5. It results in an output like Table 2. The result for posting the respective query for the restriction on the attribute *a2* is shown in Table 3. Each of the produced tables is a separate index structure for one of the restricted attributes. *CreateIndexTables* generates all indexes The created index tables allow a fast locating of interesting objects in the processing phase which is described next.

Table 2. Similarity to value 5 in R.a1

a1	Similarity
5	1
6	0.95
4	0.8
7	0.7
...	...

Table 3. Similarity to value 'Sedan' in R.a2

a2	Similarity
Sedan	1
Coupe	0.75
Van	0.4
SUV	0.3
...	...

Look Ahead. After the generation of an index structure for each restricted attribute the lists are processed top-down, i.e. row by row from highest similarity to the lowest. In contrast to Fagin's TA [19] and BPA-2 [4] the TQQA approach does not process all lists in parallel. It maintains a separate current row-number r_i for each index i. With the help of the row-numbers TQQA looks ahead to identify the most promising index: i.e. the index i for which $i[r_i + 1]$ is highest. This look ahead technique is a heuristic and has proven its applicability in [12].

Fetching Objects. The following piece of pseudo-code sketches the fetching of objects which seem to be most relevant among the remaining objects.

```
program processNext (IN string tableName,
                     IN Set<LookupTable> idx,
                     I_O Map<Number, object> objects)
   var maxIdx: LookupTable
   var q: Query
   var obj: object
   var score: Number;
begin
   LookAhead(idx, out maxIdx);
   q := select * from tableName
```

```
            where maxIdx.restriction.attr =
                  maxIdx.current_row.next.val;
    open q
    loop until no-more-rows
        fetch q into obj
        GetObjectScore(obj, idx, OUT score);
        objects[score] = obj;
    end-loop
end.
```

Within *processNext* TQQA does a look-ahead to find the most promising index avoiding fetching objects of indexes with very low similarity values. The function *processNext* is called repeatedly from within function *Top-QK* and always chooses the index structure with the highest similarity value to retrieve the next objects. Assuming the Tables 2 and 3 as the lookup structures TQQA searches for objects with the following attributes:

- Call-1: value 5 on attribute R.a1
- Call-2: value 'Sedan' on attribute R.a2
- Call-3: value 6 on attribute R.a1
- Call-4: value 4 on attribute R.a1
- Call-5: value 'Coupe' on attribute R.a2
- ...

The most promising index is used to build a query. The next objects are fetched with this query, their total scores are computed and they are added to a buffer.

Publishing Objects. This method deals with the four different combinations of top-k and top-q. It tests for each object fetched in *processNext* whether to publish it or not - depending on the score of the object, the chosen *StoppingType*, and the number of already published objects. One of the two following restrictions must hold that an object can get published:

- *at most k* objects: the object's score is at least q-% of the maximum score *and* the amount of already published objects is smaller than k.
- *at least k* objects: the object's score is at least q-% of the maximum score *or* the amount of already published objects is smaller than k.

```
program publishObjects (IN Map<Number, object> buf,
                        IN Number minScore, IN Number K,
                        IN StoppingType type,
                        I_O Map<Number, object> objects)
    var simList: List<Number>;
    var similarity: Number;
begin
    foreach Element e in buf.Elements
```

```
            if type = AT_MOST AND
                (e.score >= minScore AND objects.Count < K)
            then
                 objects[e.Score] = e.Object;
            elsif type = AT_LEAST AND
                (e.score >= minScore OR objects.Count < K)
            then
                 objects[e.Score] = e.Object;
            end-if
        end-for
end.
```

Calculating the Score of Objects. TQQA calls the objective function with a list of similarity values for a certain object from the lookup tables.

```
program GetObjectScore (IN Object obj,
                        IN Set<LookupTable> idx,
                        OUT Number score)
    var simList: List<Number>;
    var similarity: Number;
begin
    foreach Attribute a in obj.RestrictedAttrs
        similarity := idx[a][a.val];
        simList.Add(similarity);
    end-for
    score := Objective(simList);
end.
```

The method *GetObjectScore* is applied three times within the TQQA approach: (1) to calculate the maximum possible score in the initialization phase. (2) to calculate the score of each fetched object within *processNext*, and (3) to check whether to continue the search loop or not. TQQA creates the best possibly remaining object out of the current rows from the lookup tables. If the similarity of this object falls below the similarity threshold, then *contSearch(...)* terminates.

Terminating the Search. The end of the search process of $TQQA$ depends on the chosen *minimum similarity q*, the amount k, and the *StoppingType*:

- *at most k* objects: $TQQA$ *stops*, when no object can exceed q-% of the maximum score *or* when already the k objects have been published.
- *at least k* objects: $TQQA$ *stops*, when no object can exceed q-% of the maximum score *and* when at least k objects have been published.

To test these two stopping conditions $TQQA$ must check two conditions. We only have to consider objects that could be returned from the lookup tables from the current rows downwards. We create a hypothetical best matching object that could still appear and calculate its score. This hypothetical object is an object composed from all values from the current rows in the lookup tables. Table 4 and Table 5 highlight the current rows in each index after *Call-4* of *processNext*. To obtain the maximum possible score of all unseen objects a hypothetical object with values *7* on R.a1 and *Coupe* on R.a2 is created. After that the score of this object is calculated with *GetObjectScore*. With this hypothetical object $TQQA$ can determine whether there might be an object that can make it in the set of top-k objects or that can achieve a score which is at least q-% similar to the searched object.

Table 4. Index for restriction on R.a1

Current row	a1	Similarity
	5	1
	6	0.95
→	4	0.8
	7	0.7
	…	…

Table 5. Index for restriction on R.a2

Current row	a2	Similarity
→	Sedan	1
	Coupe	0.75
	Van	0.4
	SUV	0.3
	…	…

4 Distributed Top-k and Top-q Queries

4.1 Principles of the ADiT Approach

Processing a top-k query involves sending a top-k query to each peer. Optimization means determining a proper k_p for each p of the peers, i.e. how many objects should be fetched from which peer. If this k_p is too large, it results in unnecessary computations at the peers' site and unnecessary traffic. If it is too low, it is necessary to send additional queries to (some of the) peers, increasing query response time and creating higher overhead. Processing a top-q query in a peer-to-peer network with horizontal partitioning requires to send the same query with the same q to all peers and collect the results. This is fairly straight-forward and hence in the following we focus on optimizing top-k queries keeping in mind that they always can be extended with a top-q condition.

In our approach several parameters are used to calculate a k_p:

- size of the peer to peer network
- amount k of searched objects
- network capabilities of each peer, i.e. the transmission rate
- amount of objects stored on each connected peer
- speed of a peer, i.e. the searching performance of that peer.

Our approach, Adaptive Distributed Top-K query processing (short ADiT), is able to process distributed top-k queries over horizontally partitioned data exactly but the processing of the query takes the above mentioned character-istics into account. ADiT assumes a dynamic p2p network, where each peer has variable bandwidth capabilities and individual message costs. In contrast to other approaches such as [38,40] ADiT does not rely on caching techniques. Thus the performance is not dependent on stable data or on reoccurring queries.

The aim of ADiT is to achieve a low overall system effort as well as a fast query response time. The first parameter, the overall system effort is defined as the sum of all amounts of time of the peers needed for (1) sending requests to other peers in the network to obtain further objects, (2) searching objects and (3) transmitting objects. The second parameter is the query response time, the time elapsed between submitting a query and the return of the result. Formula (1) and formula (2) define the system effort, respectively the query response time where $MsgCount_i$ is the total amount of messages sent to peer P_i and n_i is the amount of objects retrieved from peer P_i. We use the following abbreviations throughout of this paper: N is the peer to peer network, Q is the top-k query, R is the queried relation, and P_i is a peer in the peer to peer system.

$$SE(N, Q, R) = \sum_{i=1}^{|P|} CCN.P_i, MsgC_i) + \tag{1}$$
$$DBCosts(N.P_i, Q, R, n_i) +$$
$$TransCosts(N.P_i, R, n_i)$$

$$QueryAnswerTime(N, Q, R) = max(CommCosts(N.P_i, MsgCount_i), \tag{2}$$
$$DBCosts(N.P_i, Q, R, n_i),$$
$$TransCosts(N.P_i, R, n_i))$$

The unit of system effort as well as of query response time is seconds. Thus it is needed to map the different costs to a time factor. Function 3 defines how sending $MsgCount$ requests to peer P is mapped to a time factor. The amount of incoming messages is multiplied with the constant costs that arise when establishing a connection to peer P. This gives the amount of time that is spent by sending $MsgCount$ messages to peer P.

$$CommCosts(P, MsgCount) = P_{MsgCosts} * MsgCount \qquad (3)$$

Function 4 defines how retrieving n objects from relation R of peer P is mapped to a time factor. The transmission costs are influenced by the size of the object in relation R on peer P and by the transmission rate of peer P.

$$TransCosts(P, R, n) = \frac{(P_{R_{ObjectSize}} * n)}{P_{TransRate}} \qquad (4)$$

The database costs ($DBCosts(N.P_i, Q, R, n)$) for searching the best n objects in relation R on peer P_i strongly depend on the top-k approach used on peer P_i, performance of the answering peer P_i, and the issued query Q, e.g. on the number of restrictions. ADiT assumes that each peer provides an estimate of the time needed to return the top-k objects for a query with m restrictions on a relation with size N. There is no assumption which procedure a peer uses to process top-k queries.

ADiT works iteratively and calculates a separate fetch size k_p' for each peer in each iteration. Then ADiT broadcasts the query Q *in parallel* and gathers the top-k_p' from each peer p. Then ADiT tries to publish objects and repeats if necessary.

There are two major possibilities for tuning: Choosing an appropriate fetch size k_p' for each peer in each iteration and avoiding to contact peers which cannot contribute to the result. For choosing the fetch size there are two extreme cases:

1. $k_p' = 1$ for each peer leads to a minimal amount of *transmitted objects* but to a higher amount of *transmitted messages*.
2. $k_p' = k$ for each peer leads to a minimal amount of *transmitted messages* but to a higher amount of *transmitted objects*.

In the rest of this paper we will focus on how to tune this basic distributed top-k query processing approach. Focusing on top-k allows easier comparison with other approaches. Extending the approach to also include top-q in the form of $TQQA$ (Sect. 3) is straightforward.

4.2 Heuristic Fetch Size Calculations

We analyzed a large number of queries by varying the influencing factors to developed two heuristics (basic and enhanced) for choosing a good fetch size k_p' for each individual peer p.

Basic Heuristics. The basic heuristics shown in Eq. 5 only uses the amount of relevant peers N_{Size} and the amount of searched objects k to derive a common fetch size f for all peers. The basic heuristics does not assume any particular data distribution. Thus it tries to retrieve an equal amount of objects from each peer. If k is larger than N_{Size}, the basic heuristics equally distributes k among the available peers. Otherwise the basic heuristics calculates the smallest multiple of k which is greater or equal than N_{Size} and equally distributes this amount among the available peers. The $consFactor$ is used to increase the fetch size since it is unlikely that each peer will contribute the same number of objects. This increasing is used to fetch more objects and keep the number of iterations small. Our initial experiments showed that a $consFactor$ of 2 leads to good results, e.g. few iterations and thus few messages exchanged in the p2p network. If the data is not distributed equally, a higher value for $consFactor$ should be chosen.

$$f = min(k, consFactor * \left\lceil \frac{N_{Size}}{k} \right\rceil * \frac{k}{N_{Size}}) \tag{5}$$

Enhanced Heuristics. The enhanced heuristics calculates the fetch size k'_p for each peer p *separately*. It uses additional parameters to adjust the fetch size for each peer properly:

- $ObjectsStored_p$: Amount of objects stored on peer p.
- $ObjectsStored_N$: Amount of objects stored in the peer to peer system N, i.e. $sum(ObjectsStored_p)$.
- $Speed_p$: Query processing speed of peer p, e.g. a value between 1 and 10 where 1 is the slowest and 10 the fastest speed.
- $maxSpeed_N$: Maximum peer query processing speed in the p2p system N.
- $TransRate_p$: The transmission rate describing how fast the network connection of a certain peer is. This value is given in MBit per second.
- $maxTransRate_N$: Maximum transmission rate in the p2p system N.

The following parameters are gathered during query processing iterations:

- $ObjectsRetrieved_p$: Amount of objects of peer p, which have already been retrieved, initially 0.
- $ObjectsPublished_p$: Amount of objects of peer p, which made it in the top-k answers, initially 0.
- $ObjPub_N$: Amount of objects returned to the user, initially 0.

All these parameters are used to calculate different weights which influence the enhanced heuristics. Applying the basic heuristics to the large test scenarios showed that the proposed fetch size should be treated as a lower limit. Therefore, the enhanced heuristics uses the different weights to *increase the fetch size* determined with the basic heuristics. The enhanced heuristics maps its weights to the interval of [1, 2] preventing to fetch fewer objects than the basic heuristics suggested. The enhanced heuristics assumes that peers that contributed more objects in previous iterations will also contribute more objects in the following iterations. This assumption is reflected in weight w_{pF} which is defined in Eq. 6. The more objects a peer published, the more objects are gathered from this peer *in the next iteration.*

$$w_{pF} = (1 + \frac{ObjectsPublished_p}{ObjPub_N}) \tag{6}$$

The enhanced heuristics tries to reduce the amount of unnecessarily fetched objects by fetching more objects from peers with a high ratio between fetched objects and published objects. Equation 7 shows the definition of weight w_{uF}.

$$w_{uF} = (1 + \frac{ObjectsPublished_p}{ObjectsRetrieved_p}) \tag{7}$$

The enhanced heuristics assumes that peers which store more objects will contribute more to the final answer. Thus it suggests to fetch more objects from larger peers. It uses Eq. 8 to incorporate that fact, namely weight w_{DBF}.

$$w_{DBF} = (1 + \frac{ObjectsStored_p}{ObjectsStored_N}) \tag{8}$$

Since it is cheap to ask a faster peer for more objects the enhanced heuristics defines w_{Speed} and $w_{TransRate}$. Equation 9 models that more objects should be fetched from peers which are faster in searching their databases.

$$w_{Speed} = (1 + \frac{Speed_p}{maxSpeed_N}) \tag{9}$$

Equation 10 deals with the transmission of objects. It reflects that more objects should be fetched from peers which have a higher transmission rate.

$$w_{TransRate} = (1 + \frac{TransRate_p}{maxTransRate_N}) \tag{10}$$

The weights described in Eqs. 6–10 are used by the enhanced heuristics to influence the basic heuristics. The weighted fetch size is determined with the heuristic function shown in Eq. 11.

$$k'_p = min(k - ObjPub_N, \lceil f * w_{pF} * w_{uF} * w_{DBF} * w_{Speed} * w_{TransRate} \rceil) \tag{11}$$

The upper bound for fetch size k'_p is obviously the amount of missing objects, namely $k - ObjPub_N$. The enhanced heuristics does not fetch more objects than the amount of missing objects from any of the peers in the peer to peer system.

4.3   ADiT Processing Iterations

ADiT processes a given distributed top-k query through a number of iterations gathering objects from the peers to answer the distributed top-k query. In this section we focus on the relevant steps in each iteration. The pseudo-code in Listing 1.1 shows how ADiT obtains the best k objects for a list of restrictions.

The variables used for storing the maximum remaining score ($maxRemScore$) and all fetched objects ($fetchedObjs$) are assumed to be globally visible to all threads during the execution. They are in-out parameters in all pseudo-codes where they are used. The output produced by the $ADiT$-method is a sorted list of the k objects which score best among all objects in the p2p system with respect to the objective function.

Identify Relevant Peers. ADiT only distributes the top-k queries to *relevant* peers. A peer p is relevant iff the last delivered object of peer p (i.e. the one with the maximum remaining score on peer p) is among the best k objects of already fetched objects, otherwise peer p is irrelevant and can be pruned, since peer p cannot return a better object than its last published object. The set of relevant peers is updated in each iteration.

Calculating Individual Fetch Sizes. In each iteration ADiT assigns an individual fetch size k'_p to each relevant peer p. The fetch size is determined using the enhanced heuristics discussed in Sect. 4.2.

```
 1 program ADiT (IN string tableName, IN Number k,
 2                 IN Set<Restriction> restr,
 3                 IN Set<Function> Sim, IN Function Obj,
 4                 I_O Map<Number, object> objects)
 5     var maxRemScore: Number;
 6     var fetchedObjs: Map<Number, object>;
 7     var ObjPublished: Number;
 8     var relPeers: Set<Peer>;
 9     var t: Thread;
10 begin
11     loop
12           maxRemScore = 0;
13           GetRelevantPeers(objects, I_O relPeers);
14           CalcFetchSize(k - objects.count, I_O relPeers);
15           -- broadcast
16           foreach Peer p in relPeers
17             t = new Thread();
18             t.start(LocalTopKCall(tableName, restr, Sim, Obj,
19                             p, fetchedObjs, maxRemScore));
20           end-for;
21           -- publish
22           PublishObjects(I_O fetchedObjs, I_O maxRemScore, k,
23                           relPeers, ObjPublished, objects);
24
25     until ObjPublished == k
26 end.
```

Listing 1.1. Pseudo-code for ADiT

Broadcasting top-k Query. Within each iteration ADiT distributes the query throughout the system and obtains k'_p objects from each peer *in parallel*.

For each relevant peer ADiT starts a separate thread (*LocalTopKCall*) which encapsulates two major tasks: (1) execution of a local top-k query and (2) updating of the maximum remaining score if it changed (Listing 1.2).

```
1  program LocalTopKCall (IN string tableName,
2                         IN Set<Restriction> restr,
3                         IN Set<Function> Sim,
4                         IN Function Obj, IN Peer p,
5                         I_O Map<Number, object> fetchedObjs,
6                         I_O Number maxRemScore)
7  begin
8      p.TQQA(tableName, q = 0, p.k', restr,
9              searchType = AT_MOST,
10             Sim, Obj, p.Objects));
11     lock(fetchedObjs, maxRemScore);
12         fetchedObjs.AddAll(p.Objects);
13         if p.maxScore > maxRemScore then
14             maxRemScore = p.maxScore;
15         end-if;
16     end-lock;
17 end.
```

Listing 1.2. Pseudo-code for sending a top-k query to a certain peer

The first part shows the call of a local $TQQA$ query processor [10] which is reentrant, i.e. gathering $k' = 5$ objects in the first iteration and $k' = 10$ objects in the second iteration finally gives the best 15 objects from peer p. After the best (or next in each following iteration) k' objects have been retrieved, they are added to a global buffer. Finally the maximum remaining score is updated if peer p has a higher maximum score than all other peers.

Publishing Objects. The last step in each iteration is the publishing of relevant objects (Listing 1.3). It is necessary to wait for all peers to return at least one result. This is indicated with the *waitForAll* method. After all peers provided their results, ADiT iterates over the *sorted* map and tests for each object whether its score is greater or equal than the maximum remaining score. In that case an object can be published. ADiT stops when enough objects have been published.

```
1  program PublishObjects (I_O Map<Number, object> fetchedObjs,
2                          I_O Number maxRemScore, IN Number k,
3                          IN Set<Peer> relPeers,
4                          I_O Map<Number, object> objects)
5  begin
6      waitForAll(relPeers);
7
8      foreach Element e in fetchedObjs.Elements
```

```
 9            if e.Score >= maxRemScore
10                objects[e.Score] = e.Object;
11
12                if objects.count == k then
13                    break;
14                end-if
15            else
16                break;
17            end-if
18        end-for
19 end.
```

Listing 1.3. Pseudo-code for the publishing of objects in ADiT

5 TQQA Prototype and Experiments

TQQA is a generic approach giving the possibility to define two functions: for measuring the similarity between values and a monotonous objective function for calculating the score of an object. In Sect. 3 we primarily focused on how these two functions are used within *TQQA* here we present potential candidates for these functions.

5.1 Similarity Function(s)

Equation 12 shows how the similarity between two attribute values is calculated. The first parameter (a) passed to *Sim* is the attribute value of a certain database tuple, the second parameter (v) is the attribute value given by the user in the restriction for an attribute. The last two parameters hold the table name t and the attribute name *attr*. This information can be used to take the data distribution of an attribute into account. Here we use the minimum and maximum values in the given table/attribute. When calculating the similarity we distinguish two different cases, e.g. the similarity between two numerical values and the similarity between two categorical values. Equation 13 shows the definition of the *NumSim(...)* used to calculate the similarity between the two given numerical values.

$$Sim(a, v, t, attr) = \begin{cases} NumSim(a, v, t, attr), & \text{if } a \text{ numerical} \\ CatSim(a, v, t, attr), & \text{otherwise} \end{cases} \tag{12}$$

$$NumSim(a, v, t, attr) = 1 - \frac{|a - v|}{max_{t_{attr}} - min_{t_{attr}}} \tag{13}$$

The numerator of Eq. 13 is the absolute difference between the two values which should be compared. The denominator models the distance between the minimum and the maximum value of an attribute in a certain relation. The farther two values are apart from each other, the smaller their similarity value is. Additionally, we take into consideration how large the distance between the minimum and the maximum value of the attribute in the given relation is. The function *NumSim* always returns a value between 0 (no similarity) and 1 (total similarity, i.e. equality). The similarity function used for categorical data is shown in Eq. 14. It returns either the inverse document frequency [7,37,39] for categorical data [2] or 0 depending whether the values are equal or not.

$$CatSim(a, v) = \begin{cases} \log \frac{n}{F(v)}, & \text{if a equals v} \\ 0, & \text{otherwise} \end{cases} \tag{14}$$

The numerator of Eq. 14 is the number of objects in a certain relation. The denominator is the frequency of value v in that given relation. Implementations of these similarity functions can be used by our $TQQA$ approach to calculate the similarity between tuples in the database and the ideal object described by a query.

5.2 Objective Function

As discussed in Sect. 3 any monotone function can serve as objective functions. Here we show how *user preference queries* can be expressed through the definition of an appropriate objective function. A user preference query allows the user to specify weights on each of the soft constraints, e.g. weights that reach from *1 (nice to have)* up to *5 (must have)*. Our example objective function, a weighted sum of similarities to the desired values is shown in Eq. 15.

$$Objective = \sum_{j=1}^{m} w_{r_{a_j}} * Sim(a_j, val_{r_{a_j}}) \tag{15}$$

Adding the possibility to specify weights on all restrictions in example ex-1 from Sect. 3.2 leads to the following two sets: S1:{(5; 3)} and S2:{(1; 1)}.

These preferences mean that the searched car *must have* 5 seats and it should be a sedan. Examining the objective function with these input parameters would lead to the value 8, the maximum achievable value for this query.

Table 6. Attribute characteristics of our test relation.

Name	Type	Selectivity
a1	Numerical	0.5
a2	Numerical	0.1
a3	Categorical	0.04
a4	Numerical	0.02
a5	Categorical	0.01
a6	Numerical	0.004
a7	Categorical	0.002
a8	Categorical	0.001
a9	Numerical	0.0001

5.3   Prototype and Experimental Setup

We implemented all algorithms described throughout this paper in a 3-tier architecture: (1) the database layer containing our TQQA approach with the described similarity (see Sect. 5.1) and objective function (see Sect. 5.2). This layer is completely embedded in the database and implemented as stored PL-SQL procedures. (It is also possible to implement it outside the database, e.g. in a special TQQA library.) (2) A database access layer which is used for sending queries to a database and receiving results. (3) A Query By Example GUI which allows easy specification of top-q queries.

The Underlying Data Model. The data model used during the experiments was a single relation containing nine attributes. In the case of having more than one relation we just create one view out of the relations involved in the query request and process the data as usual. Table 6 gives an overview on the chosen datatypes and the selectivity of each attribute. Here the selectivity is calculated as the inverse of the amount of distinct attribute values in the given relation.

To evaluate the performance of TQQA we carried out detailed measurements regarding the query response time and the memory requirements.

System Environment. All tests were carried out on a system with the following attributes: 2 Intel Xeon E5530 (Quad Core) processors each with 2.4 GHz clock rate, 16 GB RAM, Win2008 Server R2 Enterprise 64 Bit, Oracle Server 11gR2. All algorithms were implemented as stored procedures.

Test Database. The test database was filled with randomly generated data. The attributes of our test relation(s) are not correlated and their values are equally distributed. TQQA could additionally benefit from database indexes. When indexing the search attributes on database level we found that TQQA could improve its response time at an average of 1.5. Nevertheless, to test our approach and not the capabilities of the underlying database, we compared the response times without support of database indexing techniques.

Test Scenarios. For providing a meaningful analysis we adapted three parameters in our scenarios. (1) size of the relation: 10k, 20k, 50k, 100k, 200k and 500k entries, (2) minimum similarity of each returned object: we covered the interval from 10% up to 95% with a stepsize of 10%. (3) number of restricted attributes: 1 up to 9. The queries (Q1, Q2, ... Q9) posted in the test runs were constructed in the following way: Query Q1 restricts attribute a1, query Q2 additionally restricts attribute a2. Finally query Q9 restricts all available attributes and thus has the lowest selectivity of all posted queries. All these parameters were used in each possible combination for TQQA. This gives 540 queries to be executed and a profound basis to discuss the approach and draw conclusions.

Comparison. Since top-q and top-k return different amounts of objects and there is no other top-q implementation, we have chosen to compare TQQA with BAP-2 in the following way: We counted the amount of objects gathered with a specific similarity value q in a top-q query. This value was then used as amount k for searching the top-k objects with BPA-2. This allows a fair comparison because both approaches return the same amount of objects. Since typical top-k queries do not search for tens of thousands of objects we decided to start our comparison at a minimum similarity value of q = 60%. For comparing TQQA against BPA-2 we defined ratios for the query answer time (qat) and for the memory requirements:

1. $ratio_{qat} = qat_{BPA-2}/qat_{TQQA}$
2. $ratio_{mem} = bufferdObjs_{BPA-2}/bufferdObjs_{TQQA}$

For the calculation of buffered objects we counted all objects fetched from the database and stored in internal buffers. For each of the buffered objects both approaches calculated the objective function. We did not take into consideration that BPA-2 maintains one list for each restriction where *all objects* of the relation are stored. That would make the comparison even worse for BPA-2. Thus we only counted all objects examined while stepping down the sorted lists of BPA-2.

5.4   Discussion of Results

Within this section we present eight diagrams generated from the data produced by our test runs. Figure 1 shows the query answer times of TQQA on a relation with 500.000 objects and different minimum similarity values q. We observe that the query answering process gets faster the higher q is. This is expected since the approach has to examine fewer objects for a higher q. The investigation of query answer times on relations with different sizes is illustrated in Fig. 2. The query answer time grows seemingly exponential when the size of the underlying relation increases but the time needed to find interesting objects is still reasonable. In Fig. 2 we can see that it is always below 1 second when searching objects which are at least 90% similar.

In Figs. 3 and 4 we can see that the query answer time steadily increases when the selectivity of the query decreases. This finding can be observed in both figures, when we investigate different similarity values q or different relation sizes.

Additionally, we found that (1) when posting two queries with equal selectivity the query with fewer restricted attributes can be answered faster, and (2) when posting two queries with an equal amount of restricted attributes the query with the higher selectivity can be answered faster.

In Fig. 5 we can see that TQQA is between 10 and 160 times faster than BPA-2. We observe that the ratio is heavily increasing the higher q gets as fewer objects are returned for higher q. Fewer objects in the result set often means that we have to post fewer queries. This can also be observed at the rapid increase of the ratio between $q = 80\%$ and $q = 90\%$. The increase stems from a situation where TQQA only needs a few steps down in the most interesting indexes to locate the top-q objects. Fewer stepping in the indexes means fewer queries and thus results in faster query answer times. Finally when searching the top-500.000 objects both algorithms perform equal, but for a reasonable amount of objects TQQA is much faster as BPA-2.

In Fig. 6 we focus on the impact of the query selectivity. We can see that $ratio_{qat}$ is higher the higher the query selectivity is. This is because more restricted attributes often mean more queries against the database in different indexes. Those queries are time consuming and thus slow down the query answering process. The further we can observe that TQQA is more robust against the selectivity of a query the higher q is. For a similarity value $q = 90\%$ or $q = 95\%$ the $ratio_{qat}$ is almost constant.

In Figs. 7 and 8 we finally take a look at the memory ratios between TQQA and BPA-2. The scales of both figures are logarithmic at the base of 2. In Fig. 7 we can see that the memory ratio is almost between two and six. The rapid increase at the similarity value $q = 90\%$ stems from the situation that TQQA quickly found all interesting objects and did not need to further step down the index. On the other hand BPA-2 did not skip the searching process that quickly and thus examined much more objects than needed. In Fig. 8 we observe that the ratio is almost between two and five and that it is very robust against the query selectivity. The outlier at the similarity value $q = 95\%$ is due to the fact that TQQA could determine very fast that no more objects are available while BPA-2 had to examine much more objects.

In summary, TQQA was faster for *all test cases* by one to two orders of magnitude and also had much smaller memory requirements than BPA-2.

6   ADiT Prototype and Experiments

ADiT has been completely implemented in PL-SQL [22] as a set of stored procedures [36]. To compare ADiT against a state of the art distributed top-k query processing technique we also implemented the *algorithm with remainder top-k queries (short ARTO)* [38] in this database layer.

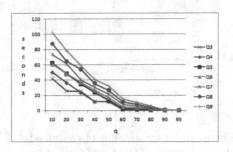

Fig. 1. Query answer times of *TQQA* on a relation with 500k objects and varying minimum similarity q.

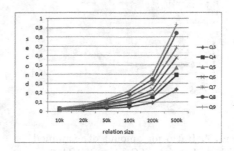

Fig. 2. Query answer times of *TQQA* for searching the $\geq 90\%$ similar objects in relations with different size.

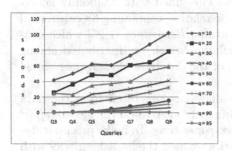

Fig. 3. Query answer times of *TQQA* on a relation with 500k objects and varying query selectivity.

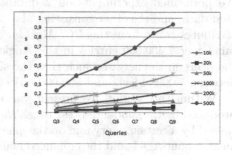

Fig. 4. Query answer times of *TQQA* for searching the $\geq 90\%$ similar objects with varying query selectivity.

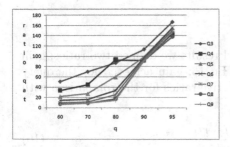

Fig. 5. Ratio of query answer times for *TQQA and BPA-2* on a relation with 500k objects and varying minimum similarity q.

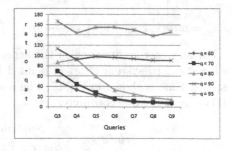

Fig. 6. Ratio of query answer times for *TQQA and BPA-2* on a relation with 500k objects and varying query selectivity.

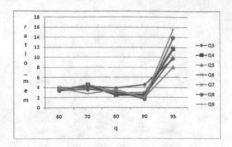 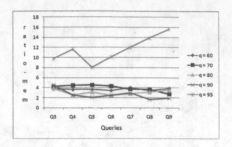

Fig. 7. Ratio for amount of examined objects between *TQQA and BPA-2* on a relation with 500k objects and varying minimum similarity q.

Fig. 8. Ratio for amount of examined objects between *TQQA and BPA-2* on a relation with 500k objects and varying query selectivity.

6.1 Experimental Setup

We performed experiments on 2 databases: One filled with randomly generated data, and the other consisting of a single relation containing 68 categorical attributes taken from the *UCI Machine Learning Repository* [1,23] which contains over 2.400.000 entries in this single relation which we distributed among the peers in the network such that the size of the database of each peer varied between 5.000 objects and 500.000 objects.

Within this section we present various diagrams generated from the data produced by the conducted test runs. We primarily focused on the *system effort* caused by a certain query and on the *query response time*. To make precise statements about ADiT and the enhanced heuristics we used the basic heuristics with a *consFactor* of 2 and four other heuristics to compare them to the enhanced heuristics:

1. $k'_p = k$
2. $k'_p = 1$
3. $k'_p = \lceil \frac{k}{N} \rceil$
4. $k'_p = \lfloor \frac{k}{N} \rfloor$
5. $k'_p = min(k, 2 * \lceil \frac{N_{Size}}{k} \rceil * \frac{k}{N_{Size}})$.

For an easier comparison of the achieved results we defined two ratios: gain with respect to system effort is defined in Eq. 16; gain achieved for the query response time is shown in Eq. 17. The respective ratios for the comparison with ARTO are defined accordingly.

$$Ratio_{SE} = \frac{SystemEffort_{heuristic_i}}{SystemEffort_{heuristic_{enhanced}}} \tag{16}$$

$$Ratio_{QAT} = \frac{QueryAnswerTime_{heuristic_i}}{QueryAnswerTime_{heuristic_{enhanced}}} \tag{17}$$

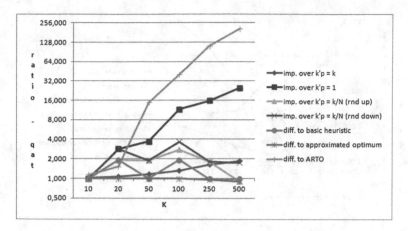

Fig. 9. Ratio for query response time between *enhanced heuristics*, approximated optimum, ARTO and five different approaches to determine the fetch size k'_p in a peer to peer system with 19 peers and varying search amount K and 4 restrictions on census data.

6.2 Discussion of Results

In Fig. 9 we can see the $Ratio_{QAT}$ for a query with 4 restrictions. Comparing with Fig. 11 we can see that all curves get higher in a peer to peer network with 49 peers. Additionally, these first figures already show that the heuristics $k'_p = 1$ is not a good choice since it involves high interaction between the query initiator and the other peers. We can also observe that for the query response time the gain over ARTO is rapidly increasing when the amount of searched objects increases. The ratio is growing fast because ARTO needs more sequential message processing when the search amount increases (when the first parallel call was not sufficient).

In Figs. 10 and 11 we can see $Ratio_{SE}$ and $Ratio_{QAT}$ for a query with 4 restrictions in a peer to peer network with 49 peers storing census data. We can observe that the ratio $Ratio_{SE}$ and $Ratio_{QAT}$ are almost identical with respect to their curves. They only differ in the magnitude which is a little higher for the $Ratio_{QAT}$. This means that the usage of ADiT brings slightly more benefits to a single user than to the whole peer to peer system. This result can be observed over all of the tests. The reason for this behaviour is that ADiT tries to fetch fewer objects from less important peers. Thus these peers do not influence the search process that much than in a setting where all peers are contributing the same amount of objects. Another reason is that the search time is dominated by the slowest peer. Avoiding high interaction and fetching few objects from such peers can clearly boost query processing.

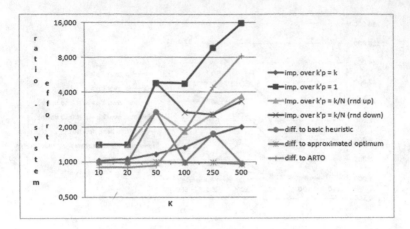

Fig. 10. Ratio for system effort between *enhanced heuristics*, approximated optimum, ARTO and five different approaches to determine the fetch size k'_p in a peer to peer system with 49 peers and varying search amount K and 4 restrictions on census data.

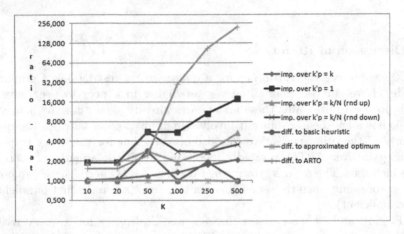

Fig. 11. Ratio for query response time between *enhanced heuristics*, approximated optimum, ARTO and five different approaches to determine the fetch size k'_p in a peer to peer system with 49 peers and varying search amount K and 4 restrictions on census data.

We observe also that ARTO has a lower system effort for a small search amount as seen in Fig. 10. The reason for this is that ARTO can answer queries with fewer messages and fewer transmitted objects. This is because ARTO sequentially asks the peer with the highest remaining score for further objects resulting in fewer work for the remaining peers. However, in Fig. 11 we can observe that the query response time is better for ADiT in the same scenario.

In Figs. 12 and 13 we can see the $Ratio_{SE}$ and the $Ratio_{QAT}$ for a query with 12 restrictions in a peer to peer network with 49 peers storing census data. In these two figures we can observe the situation where the enhanced heuristics needs more iterations than the heuristics fetching $k'_p = k$ objects. This situation only occurred once in all of the test cases. Additionally, we see the same effect as in Figs. 10 and 11, i.e. the curves are very similar but the $Ratio_{QAT}$ is a little higher than $Ratio_{SE}$.

When comparing Figs. 12 and 13 with Figs. 10 and 11 we can observe that the magnitude of the ratios is almost independent of the amount of restrictions. Furthermore, we observe in Figs. 12 and 13 that the ratios for ARTO increases at the point where the search amount exceeds the amount of peers in the network. This shows that it is better to ask each peer for more than only one object even when calling them sequentially.

The most important observations gathered through the performed test runs on random data and US Census data are:

1. ADiT is up to 200 times faster than ARTO in case the search amount gets higher than the amount of peers in the network.
2. The system effort caused by ADiT is up to 8 times lower than the system effort caused by ARTO in case the search amount gets higher than the number of peers in the network.
3. The query response time of ARTO is in most cases worse than the query response time achieved with any of the presented ADiT heuristics.

Additionally, we found some characteristics appearing in almost all test runs:

– The enhanced heuristics is close to the approximated optimum gathered through the extensive tests on the *US Census Data (1990) Data Set*.
– The enhanced heuristics is better than all other presented heuristics, except in one single query (see Figs. 12 and 13).
– The enhanced heuristics is between 2 and 32 times faster than the heuristics always fetching *1 object* from each peer in parallel.
– The enhanced heuristics is about 3 to 8 times faster than heuristics fetching $\left\lceil \frac{k}{N} \right\rceil$ or $\left\lfloor \frac{k}{N} \right\rfloor$ objects from each peer in parallel.
– The enhanced heuristics is between 1.5 and 2.5 times faster than the heuristics fetching k *objects* from each peer in parallel.
– The basic heuristics and the heuristics fetching k *objects* from each peer in parallel turned out to be better than the other heuristics.
– Heuristics fetching more objects from each peer perform better than heuristics fetching fewer objects. This can be seen when comparing the heuristics fetching $\left\lceil \frac{k}{N} \right\rceil$ vs. $\left\lfloor \frac{k}{N} \right\rfloor$ objects in parallel from each peer. The fetch sizes only differ by one, but the presented ratios show that this small difference might have a high influence on system effort and query response time.

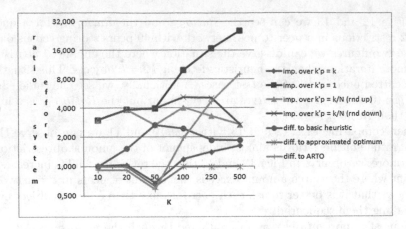

Fig. 12. Ratio for system effort between *enhanced heuristics*, approximated optimum, ARTO and five different approaches to determine the fetch size k'_p in a peer to peer system with 49 peers and varying search amount K and 12 restrictions on census data.

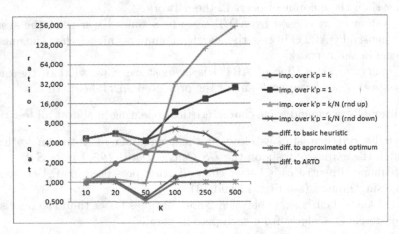

Fig. 13. Ratio for query response time between *enhanced heuristics*, approximated optimum, ARTO and five different approaches to determine the fetch size k'_p in a peer to peer system with 49 peers and varying search amount K and 12 restrictions on census data.

7 Conclusion

The goal of this work is to reduce the effort for users to retrieve relevant information from large databases where they typically face the recall/precision problem. The observed drawbacks of plain top-k queries motivated a new top-q approach called TQQA. With top-q queries users can identify objects which are at least q-% similar to the object described through the query. This allows to state a minimum degree of similarity, which is not possible with top-k approaches.

We showed that advanced indexing techniques can be used to speed up the computation of the top-q tuples. The presented technique can also be used to accelerate other types of queries with restrictions on monotone functions [16]. Finally, we compared TQQA with the best position algorithm BPA-2 [4] in a series of experiments and showed the feasibility of the approach. To the best of our knowledge BPA-2 is the fastest threshold based top-k query answering approach. We could increase query answering speed by a significant factor of 10 to 160. Additionally, our algorithm proved to have much smaller memory requirements compared to BPA-2.

We motivated the need for an adaptive approach for processing top-k queries in a distributed peer-to-peer environment. Based on data gathered through extensive experiments we derived heuristics for computing how many objects each peer should return. We evaluated these heuristics extensively in comparison with other approaches and could show that ADiT can accelerate the query response time and reduce the consumption of system resources significantly. Furthermore, we saw that the *enhanced heuristics* is in most cases close to the best system effort and query response time determined upfront. Additionally, we found that a heuristics fetching more objects is usually the better choice since returning a few more objects has much lower costs than sending an additional request. Last but not least the gains achieved with ADiT increase with the size of the peer to peer network and the number of requested results k.

References

1. UCI Machine Learning Repository, US Census Data 1990 (2012). http://archive.ics.uci.edu/ml/datasets/US+Census+Data+(1990)
2. Agrawal, S., Chaudhuri, S.: Automated ranking of database query results. In: CIDR, pp. 888–899 (2003)
3. Akbarinia, R., Pacitti, E., Valduriez, P.: Reducing network traffic in unstructured P2P systems using top-k queries. Distrib. Parallel Databases **19**, 67–86 (2006)
4. Akbarinia, R., Pacitti, E., Valduriez, P.: Best position algorithms for top-k queries. In: Proceedings of the 33rd Internatinal Conference on Very Large Databases, pp. 495–506. VLDB Endowment (2007)
5. Asslaber, M., Abuja, P., et al.: The Genome Austria Tissue Bank (GATIB). Pathobiology **74**, 251–258 (2007)
6. Balke, W.-T., Nejdl, W., Siberski, W., Thaden, U.: Progressive distributed top-k retrieval in peer-to-peer networks. In: Proceedings of the 21st International Conference on Data Engineering, ICDE 2005, pp. 174–185. IEEE Computer Society (2005)
7. Church, K., Gale, W.: Inverse document frequency (IDF): a measure of deviations from Poisson. In: Armstrong, S., Church, K., Isabelle, P., Manzi, S., Tzoukermann, E., Yarowsky, D. (eds.) Natural Language Processing Using Very Large Corpora. Text, Speech and Language Technology, vol. 11, pp. 283–295. Springer, Dordrecht (1999). https://doi.org/10.1007/978-94-017-2390-9_18

8. Ciglic, M., Eder, J., Koncilia, C.: Anonymization of data sets with NULL values. In: Hameurlain, A., Küng, J., Wagner, R., Decker, H., Lhotska, L., Link, S. (eds.) Transactions on Large-Scale Data- and Knowledge-Centered Systems XXIV. LNCS, vol. 9510, pp. 193–220. Springer, Heidelberg (2016). https://doi.org/10.1007/978-3-662-49214-7_7

9. Conner, W., Hwang, S.-W., Nahrstedt, K.: Unified framework for top-k query processing in peer-to-peer networks. Technical report, University of Illinois (2007)

10. Dabringer, C.: Efficient local and distributed query processing in a biomedical environment. Ph.D. thesis, Alpen Adria Universität Klagenfurt (2012)

11. Dabringer, C., Eder, J.: Efficient top-k retrieval for user preference queries. In: Proceedings of the 26th ACM Symposium on Applied Computing (2011)

12. Dabringer, C., Eder, J.: Fast top-k query answering. In: Proceedings of the 22th International Conference on Database and Expert Systems Applications (2011)

13. Dabringer, C., Eder, J.: Towards adaptive distributed top-k query processing. In: Ivanović, M., et al. (eds.) ADBIS 2016. CCIS, vol. 637, pp. 37–44. Springer, Cham (2016). https://doi.org/10.1007/978-3-319-44066-8_4

14. Dabringer, C., Eder, J.: Fast top-Q and top-K query answering. In: Dang, T.K., Wagner, R., Küng, J., Thoai, N., Takizawa, M., Neuhold, E.J. (eds.) FDSE 2017. LNCS, vol. 10646, pp. 43–63. Springer, Cham (2017). https://doi.org/10.1007/978-3-319-70004-5_3

15. Eder, J., Dabringer, C., Schicho, M., Stark, K.: Information systems for federated biobanks. In: Hameurlain, A., Küng, J., Wagner, R. (eds.) Transactions on Large-Scale Data- and Knowledge-Centered Systems I. LNCS, vol. 5740, pp. 156–190. Springer, Heidelberg (2009). https://doi.org/10.1007/978-3-642-03722-1_7

16. Eder, J., Frank, H., Liebhart, W.: Optimization of object-oriented queries by inverse methods. In: Eder, J., Kalinichenko, L.A. (eds.) East/West Database Workshop. Springer, LondonI (1995). https://doi.org/10.1007/978-1-4471-3577-7_8

17. Eder, J., Gottweis, H., Zatloukal, K.: IT solutions for privacy protection in biobanking. Publ. Health Genom. **15**(5), 254–262 (2012)

18. Eder, J., Koncilia, C., Morzy, T.: A model for a temporal data warehouse. In: Open Enterprise Solutions: Systems, Experiences and Organizations (OES-SEO 2001). Luiss Edizioni (2001)

19. Fagin, R., Lotem, A., Naor, M.: Optimal aggregation algorithms for middleware. In: Proceedings of the 2001 ACM Symposium on Principles of Database Systems, pp. 102–113. ACM, New York (2001)

20. Fang, Q., Yang, G.: Efficient top-k query processing algorithms in highly distributed environments. J. Comput. **9**(9), 2000–2006 (2014)

21. Fang, Q., Zhao, Y., Yang, G., Wang, B., Zheng, W.: Best position algorithms for top-k query processing in highly distributed environments. In: Proceedings of the 2010 First International Conference on Networking and Distributed Computing, ICNDC 2010, pp. 397–401. IEEE Computer Society, Washington, DC (2010)

22. Feuerstein, S., Pribyl, B.: Oracle PL/SQL Programming, 5th edn. Paperback, Sebastopol (2009)

23. Frank, A., Asuncion, A.: UCI Machine Learning Repository (2010)

24. Guntzer, U., Balke, W.-T., Kiessling, W.: Optimizing multi-feature queries for image databases. In: Proceedings of the 26th International Conference on Very Large Databases, pp. 419–428. Morgan Kaufmann Publishers Inc., San Francisco (2000)

25. Guntzer, U., Balke, W.-T., Kiessling, W.: Towards efficient multi-feature queries in heterogeneous environments. In: Proceedings of the IEEE International Conference on IT: Coding and Computing, pp. 622–628 (2001)

26. Hagihara, R., Shinohara, M., Hara, T., Nishio, S.: A message processing method for top-k query for traffic reduction in ad hoc networks. In: Proceedings of the Tenth Interenational Conference on Mobile Data Management, MDM 2009, pp. 11–20. IEEE Computer Society (2009)
27. Hofer-Picout, P., et al.: Conception and implementation of an Austrian biobank directory integration framework. Biopreservation Biobanking 15(4), 332–340 (2017)
28. Hristidis, V., Hu, Y., Ipeirotis, P.G.: Ranked queries over sources with Boolean query interfaces without ranking support. In: 26th IEEE International Conference on Data Engineering (2010)
29. Hua, M., Pei, J., Fu, A.W.C., Lin, X., Leung, H.-F.: Efficiently answering top-k typicality queries on large databases. In: Proceedings of the 33rd Interenational Conference on Very Large Databases, pp. 890–901. VLDB Endowment (2007)
30. Ilyas, I.F., Beskales, G., Soliman, M.A.: A survey of top-k query processing techniques in relational database systems. ACM Comput. Surv. 40(4), 1–58 (2008)
31. Levandoski, J.J., Mokbel, M.F., Khalefa, M.E., Korukanti, V.R.: FlexPref: a framework for extensible preference evaluation in database systems. In: ICDE, New York, NY, USA (2010)
32. Litton, J.-E.: Launch of an infrastructure for health research: BBMRI-ERIC. Biopreservation Biobanking 16, 233–241 (2018)
33. Mamoulis, N., Yiu, M.L., Cheng, K.H., Cheung, D.W.: Efficient top-k aggregation of ranked inputs. ACM Trans. Database Syst. 32(3), 19 (2007)
34. Marian, A., Bruno, N., Gravano, L.: Evaluating top-k queries over web-accessible databases. ACM Trans. Database Syst. 29(2), 319–362 (2004)
35. Nepal, S., Ramakrishna, M.: Query processing issues in image (multimedia) databases. In: ICDE, pp. 22–29 (1999)
36. Owens, K.T.: Building Intelligent Databases with Oracle PL/SQL, Triggers, and Stored Procedures, 2nd edn. Prentice-Hall Inc., Upper Saddle River (1998)
37. Robertson, S.: Understanding inverse document frequency: on theoretical arguments for idf. J. Doc. 60, 503–520 (2004)
38. Ryeng, N.H., Vlachou, A., Doulkeridis, C., Nørvåg, K.: Efficient distributed top-k query processing with caching. In: Yu, J.X., Kim, M.H., Unland, R. (eds.) DASFAA 2011. LNCS, vol. 6588, pp. 280–295. Springer, Heidelberg (2011). https://doi.org/10.1007/978-3-642-20152-3_21
39. Sparck Jones, K.: A statistical interpretation of term specificity and its application in retrieval. In: Willett, P. (ed.) Document Retrieval Systems, pp. 132–142. Taylor Graham Publishing, London (1988). http://dl.acm.org/citation.cfm?id=106765.106782. ISBN 0-947568-21-2
40. Vlachou, A., Doulkeridis, C., Nørvåg, K., Vazirgiannis, M.: On efficient top-k query processing in highly distributed environments. In: Procedings of the 2008 ACM SIGMOD International Conference on Management of Data, SIGMOD 2008, pp. 753–764. ACM (2008)
41. Wichmann, H.-E., Kuhn, K., et al.: Comprehensive catalog of European biobanks. Nat. Biotechnol. 29(9), 795–797 (2011)

Invariant Properties and Bounds
on a Finite Time Consensus Algorithm

Michel Toulouse[1]([⊠]), Bùi Quang Minh[1], and Quang Tran Minh[2]

[1] Department of Computer Science, Vietnamese-German University,
Binh Duong New City, Vietnam
`michel.toulouse@vgu.edu.vn`, `minhcly95@outlook.com`
[2] Ho Chi Minh City University of Technology, VNU-HCM,
Ho Chi Minh City, Vietnam
`quangtran@hcmut.edu.vn`

Abstract. Finite time consensus algorithms compute consensus values
exactly and in a finite number of steps, contrasting with asymptotic consensus algorithms. In the literature, there exists few approaches deriving
finite time convergence for discrete consensus algorithms. In this paper
we focus on an analysis of finite time convergence based on the observability matrix for consensus networks. We introduce analytical results
extending the applicability of network observability theory to consensus
and other distributed algorithms. New analytical bounds on the number
of steps to compute consensus are provided as well as counterexamples
which are disproving a conjecture on the minimum of steps to compute consensus. A polynomial time algorithm is described to calculate
empirically the exact number of steps to compute consensus values. We
have implemented a consensus-based network intrusion detection system
based on the observability matrix approach of consensus networks. This
implementation validates empirically our analytical results. We also compare the performance of the finite time consensus with an implementation of the same intrusion detection system using asymptotic consensus.
Although the finite time algorithm provides exact solutions, tests show
that it needs less iterations to obtain a consensus solution.

Keywords: Consensus algorithms · Finite time convergence ·
Observability theory · Distributed computing

1 Introduction

Consensus problems arise in computer network systems where unique output
values are computed distributively from several inputs. For example, nodes in a
sensor network, given they are physically distributed, may take different measurements of a same variable such as temperature, but then make consensus
on a same output value to satisfy the requirements of some wider monitoring
system. Distributed solutions to such problems can be obtained from consensus
algorithms. Let $G = (V, E)$ be a graph representing a computer network system

© Springer-Verlag GmbH Germany, part of Springer Nature 2019
A. Hameurlain et al. (Eds): TLDKS XLI, LNCS 11390, pp. 32–58, 2019.
https://doi.org/10.1007/978-3-662-58808-6_2

(the *consensus network*), where the set of n vertices V of G stands for the network nodes and the set E of edges in G stands for the network interconnection structure. Let $x_i(0)$ be the value of the measurement taken by node i and $x(0)$ the initial state of the consensus network. Assume the consensus problem consists to compute $\frac{\sum_{i=1}^{n} x_i(0)}{n}$, this problem is called *average consensus problem*. Several distributed iterative asymptotic consensus algorithms have been proposed for this problem [1–3]. These algorithms require each node i to compute a local average as follows:

$$x_i(k+1) = W_{ii}x_i(k) + \sum_{j \in \mathcal{N}_i} W_{ij}x_j(k), k = 0, 1, 2, .. \tag{1}$$

where $\mathcal{N}_i = \{j \in V | (i, j) \in E\}$ denotes the *neighbor set* of node i which consists of nodes directly connected to i in the network, W is a weight matrix, also named *consensus matrix*. For each node in the consensus network, $\lim_{k \to \infty} x_i(k) = \frac{\sum_{i=1}^{n} x_i(0)}{n}$ provided G is connected and W is row stochastic, i.e. $\mathbf{1}^T W = \mathbf{1}^T$, where T denotes matrix transpose and $\mathbf{1}$ is the vector that contains 1 in each entry. Average consensus algorithms of the form in Eq. (1) have a few drawbacks. One is that consensus is computed asymptotically, hence it is approximated. The number of iterations could be quite large before a workable approximation can be reached, i.e. the approximation is close enough to the real consensus value for the purpose of the application. Second, the only consensus value computed with this approach is the average sum of the initial values, so if the consensus problem cannot be reduced to an average sum then this approach will not help.

There are some proposals addressing the first issue, where average consensus is computed in finite time. Many focus on continuous linear and nonlinear average consensus methods, see for example [4,5] and references therein as well as [6,7]. We do not make any further references to this literature as the present paper addresses the convergence behavior of discrete consensus algorithms. We classify finite time consensus proposals into three categories: matrix factorization [8–11], minimal polynomial of the consensus matrix [12–14] and observability matrix of the consensus network [15,16].

The matrix factorization approaches consists of a sequence of minimal length of consensus matrices which are factors of the matrix $\frac{1}{n}\mathbf{1}\mathbf{1}^T$. Let D be the length of this sequence. Then we have a sequence $W(0), W(1), \cdots, W(D-1)$ of consensus matrices which are consistent with the consensus network topology such that

$$W(D-1)W(D-2)\cdots W(0) = \frac{1}{n}\mathbf{1}\mathbf{1}^T. \tag{2}$$

Given $x(0)$, the initial state of the consensus network, the sequence $W(0), W(1), \cdots, W(D-1)$ solves the average consensus problem exactly and in finite time [8,9] if

$$W(D-1)W(D-2)\cdots W(0)x(0) = \frac{1}{n}\mathbf{1}\mathbf{1}^T x(0).$$

In [8,9], the computation of the matrix factorization is centralized. In [10,11] the matrix factorization problem is solved distributively.

The minimal polynomial approach is based on the monic polynomial p of least degree d of the consensus matrix W such that $p(W) = 0$. The degree d is the number of steps to compute exactly the consensus value. Assuming the consensus matrix W is symmetric and left stochastic (assumptions relaxing these conditions are provided in [12]) where $\lambda_1, \lambda_2, \ldots, \lambda_m$ are the eigenvalues of W, then the minimal polynomial is computed as $p = \prod_{i=1}^{m}(t - \lambda_i)^{r_i}$ where r_i is the size of the largest Jordan block of W for the corresponding eigenvalue λ_i. The coefficient $\alpha_1, \alpha_2, \ldots, \alpha_{d-1}$ of the minimal polynomial p are combined to obtain a vector S of coefficients [12]

$$
S = \begin{bmatrix} 1 \\ 1 + \alpha_d \\ 1 + \alpha_{d-1} + \alpha_d \\ \vdots \\ 1 + \sum_{j=1}^{d} \alpha_j \end{bmatrix}.
$$

Let $x_i(0), x_i(1), \ldots x_i(d-1)$ be the local averages computed and stored by node i during the iterations 0 to $d-1$ of the average consensus algorithm described in Eq. (1). The vector S provides the coefficients of a linear combination of the local average values computed during the execution of the consensus algorithm to obtain exactly the average consensus value in d steps, i.e.

$$
\frac{[x_i(d-1), x_i(d-2), \ldots x_i(0)]S}{[11\ldots1]S} = \frac{\sum_{i=n}^{n} x_i(0)}{n}.
$$

Our work focuses on the finite time convergence behavior of consensus algorithms based on the observability matrix of consensus networks. In control theory, the observability matrix can be used to infer the internal states of a system from the observation of its outputs. In the context of iterative average consensus algorithms, the system is modeled as followed:

$$
x(k+1) = Wx(k), k = 0, 1, \ldots \tag{3}
$$

where Eq. (1) is the local instantiation of this system for a given node i. In the context of observability theory, the observable outputs of system described in (3) for node i are $x_i(k)$ and $x_j(k), j \in \mathcal{N}_i$. The initial vector $x(0)$ is the internal state to infer. At each iteration of Eq. (1), node i stores $x_i(k)$ and $x_j(k), j \in \mathcal{N}_i$ as these values are directly accessible to node i. An observation matrix O_i is built for each node, the dimension of O_i determines the number of observations of $x_i(k)$ and $x_j(k), j \in \mathcal{N}_i$ required to infer $x(0)$ from the values stored by node i. This provides a finite time consensus algorithm.

The finite time consensus algorithm based on the observability matrix has first been proposed in [15,16]. So far, only lower and upper bounds on the number of observations existed in the literature. Our work provides new mathematical developments for this algorithm in terms of new invariant properties and bounds on the execution time of the algorithm, an empirical approach to compute exact bounds for any consensus network, finally counterexamples to a conjecture (conjecture 1) in [17], thus disproving the conjecture. As a second contribution, we have embedded the finite time average consensus algorithm based on the observability matrix approach in an network intrusion detection system (NIDS). This implementation is based on a previous work [18] where an asymptotic average consensus algorithm supports a distributed NIDS. We compare the performance of the two implementations and use the implementation of the finite time distributed NIDS to validate our analytical results.

Finally, for completeness, we mention game theory, which has entries in the literature on continuous and asymptotic average consensus algorithms but we are aware of only one reference related to finite time consensus [19]. Very briefly, the average consensus algorithm in [19] uses the characteristic polynomial of the matrix $(I_n + \gamma L)$ to determine an upper bound on the finite number of steps needed to compute a consensus value, where I_n is the $n \times n$ identity matrix, L is the Laplacian of the consensus matrix W and γ is a negative coefficient ensuring the eigenvalues of $(I + \gamma L)$ are smaller or equal to 1.

The present paper is organized as follow. Next section describes the consensus problem, solutions to this problem based on average consensus algorithms, and a brief introduction to observability theory and its application to distributed average consensus. Section 3 describes the analytical results in [15,16], providing definitions and an algorithmic framework for the subsequent sections. Section 4 describes our analytical results. Section 5 contains the implementation of a network intrusion detection system where aggregated information is computed using the finite time consensus algorithm described in Sect. 3. Finally Sect. 6 concludes the paper.

2   Background on Average Consensus and Observability Theory

Consensus problems appear in hundreds of applications in computer science and control theory. In this section we describe applications where these problems arise, formalize solutions to consensus problems, recall some observability theory contributions to average consensus in cyber security and motivate its application to finite time consensus.

2.1   The Consensus Problem

Consensus problems arise in distributed systems such as computer networks, sensor systems, multi-agent systems and others. Consensus problems appear more acutely when distributed systems apply redundancies in order to improve trust, reliability, robustness and security. This is the case for example in multiple sensors systems where observations are duplicated to obtain more accurate information compared to a single source observation or to use smaller and cheaper sensors for the same performance. In distributed databases, particularly blockchains, consensus is needed to ensure consistency among duplicate data stored on different servers. In air traffic control, duplicate processes help to improve security.

As a more detailed illustration, consider *cooperative spectrum sensing* in cognitive radio where robustness is obtained through redundancy. Spectrum sensing is used to identify temporarily unoccupied bandwidth frequencies that can be made available to secondary users (SUs) without interfering with their utilization by primary users (PUs) [20]. Spectrum sensing consists to observe the RF environment in order to detect whether an PU is currently using the spectrum. This detection is based on measurements of the energy level. High energy levels are indications of a busy spectrum. Spectrum sensing faces the usual performance issues associated with wireless communication caused by weather, multipath propagation or physical terrain impacting wave propagation. Cooperative spectrum sensing addresses those issues by having the identification of unoccupied frequencies performed in a spatially distributed manner by the secondary users interested in getting access to unused frequencies. It is unlikely that the measurements taken by each sensing SU will yield the same value. Therefore, a consensus value has to be computed to which the sensing devices agree on the energy level of the sensed spectrum. Several data fusion algorithms for cooperative spectrum sensing are based on consensus algorithms [21].

2.2   Formalization of Consensus Algorithms

Equation (1) is the core local linear model for discrete average consensus algorithms while Eq. (3) is an iterative model of the distributed algorithm, the network wide update rule. The non-zero entries of the corresponding weight matrix W are modeled on the adjacency structure of the graph G, i.e. $W_{ij} = 0$ if $(i, j) \notin E$, otherwise W_{ij} represents a weight on edge $(i, j) \in E$. Furthermore, W has non-zero entries for self-edges, i.e. $W_{ii} \neq 0$, these values are used to ensure consensus convergence, see [3] for the convergence conditions that W must satisfy. It is relatively easy to find a weight matrix that satisfies convergence conditions. For example, $W = I_n - \sigma L$, with $0 < \sigma < \frac{1}{\max |\mathcal{N}_i|}$ is such weight matrix where $\max |\mathcal{N}_i|$ is the neighborhood with the largest cardinality. The following weight matrix, the *Metropolis-Hasting* matrix, also satisfies the convergence conditions for solving the average sum consensus problem:

$$
W_{ij} = \begin{cases} \frac{1}{1+\max(deg_i, deg_j)} & \text{if } i \neq j \text{ and } j \in \mathcal{N}_i \\ 1 - \sum_{k \in \mathcal{N}_i} W_{ik} & \text{if } i = j \\ 0 & \text{if } i \neq j \text{ and } j \notin \mathcal{N}_i \end{cases} \tag{4}
$$

where deg_i denotes the degree of node $i \in G$.

Assuming that the system in Eq. (3) converges, and converges to a correct consensus value, a second issue is the convergence speed of the consensus algorithm. The coefficients of the weight matrix impact convergence speed [3]. Convergence speed also depends on the network topology. In this context, the graph

Laplacian, specifically its second eigenvalue, is used to analyse the convergence speed [22] of asymptotic consensus algorithms. In this work we compare asymptotic convergence behavior of consensus algorithms and their approximated solutions with finite time consensus algorithms and their exact solutions.

2.3 Observability Theory

Observability theory is a broad mathematical development of control theory. It is used for example to determine whether the internal states of a system can be inferred from its observable outputs. If so, the system is deemed "observable", algorithms can be found to infer the internal states of a system in finite time. Observability is a pre-condition to system controllability. It has several control engineering applications concerned with identifying and locating faults in a system. *Observer-based* techniques [23] based on *model-based* fault detection are such applications where the anticipated behavior of a system is described using mathematical models [24]. Control theory solutions to the identification of faults in a system, based on observer-based fault detection approaches, have been applied to design cyber-attack detection systems [25–27]. They have also been extended to detect data falsification attacks on asymptotic average consensus algorithms [28–30].

In data falsification attacks, observer-based systems detect anomalies in the execution of asymptotic consensus algorithms. In the present work, observability theory is used to analyze the core linear iterate (Eq. 1) of asymptotic consensus algorithms as an information diffusion and an encoding scheme. Like in any distributed algorithm, the execution of an asymptotic consensus algorithm causes information to flow among the nodes involved in the distributed computation. Information flows according to a pattern that depends on the distributed algorithm and the network topology. Each node i in a consensus algorithm, prior to compute $x_i(k+1)$, receives the states $x_j(k)$ for $j \in \mathcal{N}_i$ using the communication links adjacent to node i. Once state $x_i(k+1)$ is computed, this state is sent to all the neighbors of node i using the communication links adjacent to node i. Of particular interest, if the number of iterations is large enough, this information flow pattern causes the initial state of each node to reach each other node in the network.

However, consensus is not a value communication protocol. Each iteration in Eq. (1) compresses the states $x_i(k)$ and $x_j(k)$ for $j \in \mathcal{N}_i$ into a new state $x_i(k+1)$. This is analogue to network coding [31] where, at the network layer, packets from different sources are recombined into a single packet to improve network's throughput. An initial state $x_i(0)$ that reaches another node j after k consensus iterations would have been compressed k times, therefore the initial value $x_i(0)$ is not readily available to node j. As in network coding, for node j to recover the initial states of all the other nodes, it needs an inverse of the encoding scheme executed each time Eq. (1) is iterated. The original contribution in Sundaram and Hadjicostis [15,16] is to apply observability theory to the analysis of the information flow in a consensus network and to design an inverse function recovering $x(0)$ from the flow of values that pass through a given node i during the execution of a consensus algorithm.

3   A Finite Time Consensus Algorithm

The finite time consensus algorithm described in this section is based on the analytical results in [15, 16]. The algorithm we describe is a distributed algorithm in which a same procedure is executed by each node of a consensus network. That procedure is the linear iterate of Eq. (1) plus instructions implementing the observation and storage of network traffic flowing through each node, as well as code for a matrix multiplication which infers $x(0)$. We divide this section along these two main phases of the finite time consensus: the observation phase and the inference phase. We review some implementation issues and limitations of this finite time consensus algorithm. Last, it should be noted that the algorithm describes in this section (as well as the results in [15,16]) defines a procedure to compute in finite time any consensus function, where average consensus is just a special case.

3.1   Observation Phase

The setting of a consensus network, in which no explicit routing protocol is used, constraints any given node i at each iteration k to only access its own state $x_i(k)$ and the states $x_j(k)$ of its neighbors in the consensus network. These states constitute information flowing through node i, they are the observations that node i can readily make about the system in Eq. (3). In the context of the modelization proposed in [17], these observable states are the outputs for node i of the system in Eq. (3). Keeping with the model in [17], the system states $x(0), x(1), \ldots, x(k)$ are unknown in their entirety to each particular node, these states can be considered as the internal states of the system that a particular node could be interested to infer. From the perspective of consensus theory, among the internal states $x(0), x(1), \ldots, x(k)$, the internal state that each node i needs to infer is $x(0)$, the initial state of each node in the consensus network. Once $x(0)$ is known, i.e. the input parameters of a consensus function, then node i can compute directly the consensus function.

Borrowing from observer models in control theory, the observable states $x_j(k)$ of the consensus system from the perspective of node i are expressed by the following linear iterative system:

$$
\begin{aligned}
x(k+1) &= Wx(k) \\
y_i(k) &= C_i x(k)
\end{aligned}
\tag{5}
$$

where $y_i(k)$ is the output of the system[1] $Wx(k)$ observed by node i at iteration k. C_i is a $(deg_i + 1) \times n$ matrix with a single 1 in each row, indicating the values of

---

[1] In a more general set up, the system in Eq. (5) includes a control term which is used to maintain the system in a desired region of its state space. The synthesis of such controller is the subject of research related to continuous consensus which yields finite time algorithms, see for example [32,33] and references therein.

vector $x(k)$ stored in $y_i(k)$ (entry $C_i[k,j] = 1$ if $j \in \mathcal{N}_i$, otherwise $C_i[k,j] = 0$). At any given iteration k, the extraction of the vector $C_i x(k)$ causes node i to store $(deg_i + 1)$ states in y_i, i.e. the states $x_i(k)$ and $x_j(k) \in \mathcal{N}_i$ $(deg_i = |\mathcal{N}_i|)$. Notations $y_i(k-1)$ and $y_i(k)$ refer to two different sets of entries in y_i. The number of states stored in $y_i(0 : v-1)$ is $(deg_i + 1)v$, where v is the length of the sequence of observations performed by node i.

As an illustration of the system in Eq. (5), we consider the consensus network represented by the graph in Fig. 1, where nodes 1, 2, 3 and 4 are initialized respectively with values 1, 2, 3 and 4, i.e. $x(0) = [1, 2, 3, 4]$. Assume the consensus function to be computed is $f = \frac{1}{n} \sum_{i=1}^{n} x_i(0)$, the average sum function, where the consensus value is $f = \frac{1}{n} \sum_{i=1}^{4} x_i(0) = 2.5$.   Assume the coefficients of

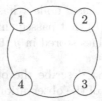

Fig. 1. Ring network

the weight matrix W are derived from the Metropolis-Hasting weight matrix as defined in Eq. (4) which, for a 4-nodes ring network, yields the following coefficients:

$$W = \begin{bmatrix} 0.33 & 0.33 & 0.00 & 0.33 \\ 0.33 & 0.33 & 0.33 & 0.00 \\ 0.00 & 0.33 & 0.33 & 0.33 \\ 0.33 & 0.00 & 0.33 & 0.33 \end{bmatrix}.$$

We would like to record the observations made by node 1 in the above network. The corresponding C_1 matrix is as follows:

$$C_1 = \begin{bmatrix} 1 & 0 & 0 & 0 \\ 0 & 1 & 0 & 0 \\ 0 & 0 & 0 & 1 \end{bmatrix}.$$

Next the entries of y_1 are computed as follows. After the first iteration of the observation phase, the content of the column vector $y_1(0)$ records the value of $x_1(0) = 1$ and the initial values of its neighbors, i.e. $x_4(0) = 4$ and $x_2(0) = 2$.

$$y_1(0) = \begin{bmatrix} 1 & 0 & 0 & 0 \\ 0 & 1 & 0 & 0 \\ 0 & 0 & 0 & 1 \end{bmatrix} \begin{bmatrix} 1 \\ 2 \\ 3 \\ 4 \end{bmatrix} = \begin{bmatrix} 1 \\ 2 \\ 4 \end{bmatrix}.$$

This corresponds to the observation made by node 1 at iteration 0. At the second iteration of the observation phase, the values stored in $y_1(1) = C_1 W x(0)$

$$y_1(1) = \begin{bmatrix} 0.33 & 0.33 & 0.00 & 0.33 \\ 0.33 & 0.33 & 0.33 & 0.00 \\ 0.33 & 0.00 & 0.33 & 0.33 \end{bmatrix} \begin{bmatrix} 1 \\ 2 \\ 3 \\ 4 \end{bmatrix} = \begin{bmatrix} 2.31 \\ 2.00 \\ 2.66 \end{bmatrix}.$$

After the execution of two consensus iterations, i.e. two observations, y_1 is the column vector representation of $[1, 2, 4, 2.31, 2, 2.66]$.

3.2 Inference Phase

At the end of the observation phase, node i has stored in $y_i(0 : v - 1)$ the values of the states $x_i(k)$ and $x_j(k)$ that passed through node i at each iteration $k = 0..v - 1$. The $(deg_i + 1)v$ values stored in $y_i(0 : v - 1)$ can be used to infer the initial state $x(0)$.

In the present section we first describe the observability matrix, which in itself is an analytical description of the observation phase. Next we show how to infer $x(0)$ from the information stored during the observation phase.

Analysis of the Observation Phase. We know from linear algebra that $x(k) = W^k x(0)$ for W^k the product of matrix W with itself k times. After the first iteration of the observation phase, node i stores $y_i(0) = C_i W^0 x(0)$. At the next iteration, $y_i(1) = C_i W x(0)$, then $y_i(2) = C_i W^2 x(0)$ and so forth until $y_i(v - 1) = C_i W^{v-1} x(0)$. The sequence of v observations performed by a given node i corresponds mathematically to the sequence $(C_i W^0 x(0), C_i W x(0), C_i W^2 x(0), \ldots, C_i W^{v-1} x(0))$. This sequence of operations in matrix form is called the *observability matrix* denoted as $O_{i,v-1}$:

$$O_{i,v-1} = \begin{bmatrix} C_i \\ C_i W \\ C_i W^2 \\ \vdots \\ C_i W^{v-1} \end{bmatrix}. \tag{6}$$

The observability matrix is the analytical form of the information diffusion and encoding processes taken place during the sequence of observations performed by node i. The observability matrix is the transformation matrix that converts the states in $x(0)$ into the data stored in $y_i(0 : v - 1)$:

$$\begin{bmatrix} y_i(0) \\ y_i(1) \\ y_i(2) \\ \vdots \\ y_i(v_i - 1) \end{bmatrix} = \begin{bmatrix} C_i \\ C_i W \\ C_i W^2 \\ \vdots \\ C_i W^{v_i-1} \end{bmatrix} x(0). \tag{7}$$

Because C_i is unique to each node in the consensus network, there exists one different observability matrix for each node. The following is the observability matrix $O_{1,1}$ for node 1 of the ring network in Fig. 1 for $v = 2$:

$$O_{1,1} = \begin{bmatrix} C_1 \\ C_1 W \end{bmatrix} = \begin{bmatrix} 1.00 & 0.00 & 0.00 & 0.00 \\ 0.00 & 1.00 & 0.00 & 0.00 \\ 0.00 & 0.00 & 0.00 & 1.00 \\ 0.33 & 0.33 & 0.00 & 0.33 \\ 0.33 & 0.33 & 0.33 & 0.00 \\ 0.33 & 0.00 & 0.33 & 0.33 \end{bmatrix}.$$

Recovering $x(0)$. Before describing a procedure to infer $x(0)$, we first describe a procedure that computes the consensus function directly from the data stored by node i in vector y_i. The procedure to infer $x(0)$ derives directly from this first procedure.

Let $f(x(0))$ be the consensus function to be computed. A function f'_i is defined for each node i such that $f'_i(y_i(0 : v - 1)) = f(x(0))$. First a matrix representation Q of f is found such that $f(x(0)) = Q(x(0))$. Assume $f = \frac{1}{n}\sum_{i=0}^{n}$. The corresponding matrix representation of f is the row vector $1 \times n$, where $Q = [\frac{1}{n}, \frac{1}{n}, \ldots]$. The function f'_i is only known in its matrix form, denoted as Γ_i, which is computed out of the observability matrix $O_{i,v-1}$. The matrix Γ_i can be recovered from $O_{i:v-1}$ provided Q, the matrix representation of f, is in the row space of $O_{i,v-1}$ (see the proof in [16]). The function Γ_i is therefore a linear combination of the rows of $O_{i,v-1}$. To obtain Γ_i, the following system of linear equations is solved:

$$O_{i,v-1}^T \Gamma_i^T = Q^T. \tag{8}$$

The function Γ_i computes the consensus function $f(x(0))$ using the values stored in $y_i(0 : v - 1)$:

$$\Gamma_i \begin{bmatrix} y_i(0) \\ y_i(1) \\ y_i(2) \\ \vdots \\ y_i(v - 1) \end{bmatrix} = \Gamma_i O_{i,v-1} x(0) = Q x(0). \tag{9}$$

Continuing our previous illustration for node 1, $v = 2$ and the observability matrix $O_{1,2}$, we solved the system $O_{1,2}^T \Gamma_1^T = Q^T$:

$$\begin{bmatrix} 1.00 & 0.00 & 0.00 & 0.33 & 0.33 & 0.33 \\ 0.00 & 1.00 & 0.00 & 0.33 & 0.33 & 0.00 \\ 0.00 & 0.00 & 0.00 & 0.00 & 0.33 & 0.33 \\ 0.00 & 0.00 & 1.00 & 0.33 & 0.00 & 0.33 \end{bmatrix} \Gamma_1^T = \begin{bmatrix} 0.25 \\ 0.25 \\ 0.25 \\ 0.25 \end{bmatrix}$$

which has infinitely many solutions, one is $\Gamma_1 = [-0.66, -0.33, -0.75,]$ $[2, -0.25, 1]$, expressing Q in terms of a set of linearly independent rows of $O_{1,2}$

as a new base for Q. Using the column form of $y_1 = [1, 2, 4, 2.31, 2, 2.66]$ that we have previously computed, it can be verified that $\Gamma_1 y_1 = 2.46 \approx 2.5$, the difference is accounted for by rounding errors.

The procedure to infer $x(0)$ consists to solve Eq. (8) for the identity matrix I_n in place of the matrix representation Q of the consensus function f. Assuming I_n is in the row space of $O_{i,v-1}$, the solution of $O_{i,v-1}^T \Gamma_i^T = I_n$ expresses I_n as a linear combination of the rows in $O_{i,v-1}$, from which $\Gamma_i y_i$ returns $x(0)$.

Using the previous illustration for computing the average sum function, first the identity matrix I_4 is expressed in a base provided by the $O_{1,2}$ matrix, by solving the system of linear equations $O_{1,2}^T \Gamma_1^T = I_4$:

$$
\begin{bmatrix}
1.00 & 0.00 & 0.00 & 0.33 & 0.33 & 0.33 \\
0.00 & 1.00 & 0.00 & 0.33 & 0.33 & 0.00 \\
0.00 & 0.00 & 0.00 & 0.00 & 0.33 & 0.33 \\
0.00 & 0.00 & 1.00 & 0.33 & 0.00 & 0.33
\end{bmatrix}
\Gamma_1^T = I_4.
$$

The solution of this system of equations is as follows:

$$
\Gamma_1^T =
\begin{bmatrix}
0.917 & -0.028 & -0.836 & -0.082 \\
-0.083 & 0.953 & -0.336 & -0.033 \\
-0.083 & -0.008 & -0.336 & 0.869 \\
0.250 & 0.083 & -0.497 & 0.249 \\
0.000 & 0.058 & 1.515 & -0.149 \\
0.000 & -0.058 & 1.515 & 0.149
\end{bmatrix}.
$$

Let $x_i(0) = [1, 2, 3, 4]$, and $y_1 = [1, 2, 4, 2.31, 2, 2.66]$. One can verify that $\Gamma_1 y_1 = [0.9965, 1.9994, 3.0598, 4.0015]$, rounding errors accounting for the differences. Node 1 has then learned the initial state of the system, therefore node 1 can compute directly the average sum using the $f = \sum_{i=1}^n x_i(0)$. Node 1 can also compute as well almost any other consensus function of $x(0)$.

3.3   Implementation Issues

The observability matrices $O_{i,v-1}$, the matrices Γ_i, as well as the matrices Q, C_i and I_n are pre-computed prior to the execution of the observation and inference phases. These matrices can be computed by a more powerful central node as, for example, the computation of Γ_i requires the execution of a linear least squares solver for computing the underdetermined systems of linear equations $O_{i,v-1}^T \Gamma_i^T = I_n$. Only matrix Γ_i has to be sent to node i of the consensus network. Algorithm 1 below summarizes the finite time consensus algorithm that has been described in this section. This algorithm is executed by each node of a consensus network.

The properties of the observability matrices are conditioned by the weight matrix W. A key condition is whether there exists a weight matrix W such that the consensus function f or the identity matrix I_n is in the row space of $O_{i,v-1}$. It turns out that all weight matrices that satisfy the convergence

Algorithm 1. Finite time consensus procedure for node i

Input: Γ_i, v, W, f
Observation phase:
for $(k = 0; k < v; k + +)$ **do**
$\quad x_i(k + 1) = W_{ii}x_i(k) + \sum_{j \in \mathcal{N}_i} W_{ij}x_j(k)$
$\quad y_i(k) = x_i(k) \cup x_j(k) \; j \in \mathcal{N}_i$
end for
Inference phase:
$x(0) = \Gamma_i y_i$
Computing the consensus value:
return $f(x(0))$

conditions of asymptotic consensus can be used to compute consensus in finite time directly from the vector y_i (see "Designing the weight matrix" in [16]). Furthermore, many weight matrices which do not depend on the magnitude of their eigenvalues, which will typically not be allowed for asymptotic consensus, do exist for finite time consensus [16]. For the general case where $x(0)$ is inferred from $y_i(0 : v - 1)$, the rank of the matrix I_n is n, therefore I_n is in the row space of $O_{i,v-1}$ only when $O_{i,v-1}$ is full column-rank. It has been proven that random weight matrices can generate a full column-rank $O_{i,v-1}$ with probability equal to 1 [17].

In order to compute the observability matrices $O_{i,v-1}$, the value of v, the length of the observation phase sequence has to be known in advance. This is a complicated issue, prior to the present work only analytical lower and upper bounds on the length of this sequence were known. The purpose of the observation phase is for each node i to record enough information such that the initial state $x(0)$ can be inferred. Network topology impacts this number of observations. A node i cannot infer the initial state $x_j(0)$ of node j unless $x_j(0)$ has propagated to node i during the observation phase, the length of the shortest path from j to i is therefore a lower bound on the length of the observation sequence for node i. For regular network topologies, a lower bound is the diameter of the consensus network. The number n of nodes in a consensus graph is an upper bound, after n iterations of the observation phase, the initial state of each node has propagated to each other node in the network. Better upper bound is $n - deg_i$ [15]. If the consensus network is not a regular graph, then v is likely to be different among the nodes in the graph, there is a number of observations v_i specific for each node i. We address those issues in the next section where we introduce an algorithm to compute empirically the exact value of v for any network topology, as well as introducing a few new analytical bounds.

4 Properties of the Finite Time Consensus Algorithm

This section introduces some new properties of the finite time consensus algorithm described in Sect. 3. We first show that the observability matrix can also be used to infer any internal state $x(k)$ from the system described in Eq. (3).

The initial state $x(0)$ can be obtained from $x(k)$ using the inverse of the weight matrix W. Next we show the non-increasing rank step property of the observability matrix $O_i(0 : v-1)$ also applies to any observability matrix $O_i(k : v-1+k)$, i.e. the start iteration of the observations has no impact on the number of iterations of the observation phase of the above finite time consensus algorithm. Later, we derive analytical bounds on the length of the observation phase and finally we introduce a polynomial time algorithm that computes the exact number of iterations of the observation phase for any network topology, and we list exact values for a set of well-known consensus network topologies.

4.1   Irrelevance of the Start Iteration

As discussed above, the finite time consensus algorithm can recover the initial state $x(0)$ from the observations of any node by setting the target function $f = I_n$. Once $x(0)$ is recovered, any state at any given iteration k can be calculated by $x(k) = W^k x(0)$. Thinking in reverse, if $x(k)$ is known, the initial state can be deduced using $x(0) = W^{-k}x(k)$, given that W is invertible. Consequently $x(k)$ could be retrieved by assuming it is the initial state and conducting the finite time consensus algorithm as normal. Given the weight matrix W is invertible (which is mostly the case), observations do not necessarily have to start from the first iteration (iteration 0), but can start from an arbitrary iteration k. If v is the number of iterations of the observation phase, the initial state $x(0)$ could be recovered by using the observations either from iteration 0 to $v - 1$ or from iteration k to $k + v - 1$. Not only $x(0)$ can be recovered, but the process of building the observability matrix O_i must be the same for any start iteration k, including the rank of each building step. For any node i, we denote:

$$O_i(p : q) = \begin{bmatrix} C_i W^p \\ C_i W^{p+1} \\ \vdots \\ C_i W^q \end{bmatrix} \tag{10}$$

the observability matrix of node i representing the observations from iteration p to iteration q of its observation phase. The irrelevance of the start iteration property could be expressed formally as:

$$\rho(O_i(0 : q)) = \rho(O_i(k : k + q)) \tag{11}$$

where $\rho(A)$ is the rank of matrix A, k is an arbitrary start iteration, and $q \geq 0$ is an arbitrary end iteration.

Proof. The two matrices are related by the formula: $O_i(k : k+q) = O_i(0 : q)W^k$. The rank of a matrix product is always less than or equal to those of its factors, hence $\rho(O_i(k : k + q)) \leq \rho(O_i(0 : q))$. However, because W is invertible, the above mapping equation could be written in reverse order: $O_i(0 : q) = O_i(k : k + q)W^{-k}$; and the inequality also happens in the other ways. Therefore, the rank of the two matrices must be equal: $\rho(O_i(k : k + q)) = \rho(O_i(0 : q))$.

4.2   Non-increasing Rank Step of the Observability Matrix

Given a start iteration k, the rank step of O_i t iterations after the start iteration, is defined as:

$$\Delta_i(t) = \rho(O_i(k : k + t)) - \rho(O_i(k : k + t - 1)) \tag{12}$$

and the rank step at the start iteration k is $\Delta_i(0) = \rho([C_i W^k]) = \deg_i + 1$. In other words, the rank step $\Delta_i(t)$ is the number of independent rows of $C_i W^{k+t}$ against $O_i(k : k+t-1)$. The start iteration k is irrelevant, because for any other start iteration l, $\rho(O_i(k : k + t)) = \rho(O_i(l : l + t))$ and $\rho(O_i(k : k + t - 1)) = \rho(O_i(l : l+t-1))$, and their difference will still be the same. The rank step $\Delta_i(t)$ is a non-increasing function given $t \geq k$. Formally, for all $t > k$:

$$\Delta_i(t) \leq \Delta_i(t - 1). \tag{13}$$

Proof. $\Delta_i(t)$ is the number of independent rows of $C_i W^{k+t}$ against $O_i(k : k+t-1)$. Removing $C_i W^k$ from $O_i(k : k+t-1)$ creates $O_i(k+1 : k+t-1)$ and frees up some dimensions in the row space of O_i, thus the number of independent rows of $C_i W^{k+t}$ will increase or at least remain the same. Because of the irrelevance of the start iteration, the number of independent rows of $C_i W^{k+t}$ against $O_i(k+1 : k + t - 1)$ is the same as that of $C_i W^{k+t-1}$ against $O_i(k : k + t - 2)$, which is in fact $\Delta_i(t - 1)$. Therefore, $\Delta_i(t - 1) \geq \Delta_i(t)$.

Note that for all $t > 0$,

$$\Delta_i(t) \leq \deg_i < \Delta_i(0). \tag{14}$$

The reason is while $C_i W^{k+t}$ has $\deg_i + 1$ rows, the rows corresponding to the observation of node i always depend on previous observations because of the definition of the consensus algorithm: $x_i(t) = W_{ii} x_i(t-1) + \sum_{j \in \mathcal{N}_i} W_{ij} x_j(t-1)$. Thus, there is only at most $\deg_i$ independent rows in $C_i W^{k+t}$. The only exception is at the first iteration, iteration 0, where the observation of node i is its initial state which does not depend on any previous observation (because there is none).

4.3   Bounds on the Iterations of the Observation Phase

Upper Bound. In the first iteration, any node i will receive the initial states of its own and its neighbor nodes, which results in a rank step $\Delta_i(0) = \deg_i + 1$. Let v_i be the iteration corresponding to the observation phase of node i, the *observability index* of node i. For each of the consecutive iterations $t < v_i$, the rank step $\Delta_i(t)$ must be at least 1 (because of the non-increasing rank step property, if $\Delta_i(t)$ falls to 0, it will never come back up, and the initial state would never be recovered). Thus, $v_i \leq n - \deg_i$ and the upper bound on the number of iterations is:

$$v \leq n - \min_i(\deg_i). \tag{15}$$

This upper bound about this finite time consensus was first published in [16]. In [17], Sundaram also derived an upper bound on v_i by partitioning the consensus graph (excluding the node i) into connected subgraphs. Each subgraph contains one and only one neighbor of node i. Then the upper bound is the minimum size of the largest subgraph. Formally, let $\mathcal{P}$ be the partitioning of G, $S_{\mathcal{P}}(j)$ where $j \in \mathcal{N}_i$ be the connected subgraph corresponding to the neighbor node j of i while in $\mathcal{P}$, then:

$$v_i \leq \min_{\mathcal{P}}(\max_j |S_{\mathcal{P}}(j)|). \tag{16}$$

Another equivalent structure to this set of subgraphs is a spanning tree rooted at i. Then each subgraph will correspond to a branch of the tree and the bound would be the minimum size of the largest branch among all the possible spanning tree rooted at node i.

Lower Bound. For any node i, there are two basic lower bounds for the observability index v_i:

- Node eccentricity. This is equivalent to the number of consensus iterations needed such that the data of the furthest node in G (with respect to the node i) could be transmitted to i. Generalized to the whole graph, the lower bound on the observation phase is the diameter of the graph.
- The quotient $(n-1)/\deg_i$. This lower bound could be obtained by assuming the rank step $\Delta_i(t)$ is always the largest possible value, i.e. $\Delta_i(k) = \deg_i$ if $t > 0$ and $\Delta_i(k) = \deg_i + 1$ if $t = 0$. This value is equivalent to the time needed to get all the data with the maximum input capacity.

The first lower bound is more accurate for graphs with a large diameter and low maximum degree such as path or ring graphs. The second one is good for graphs where the increase in the number of nodes for which node i has received data is super-linear over the iterations, like mesh or torus graphs (for torus this number increases quadratically at each observation iteration). For graphs which are hybrid between the two categories, other factors should be considered such as bottlenecks.

Bottlenecks. Bottlenecks (with respect to a node i) are nodes in a consensus graph G which limit the input capacity of a node i. In other words, they are the factors which decrease the rank step $\Delta_i(t)$. Consider a graph with a vertex cut set K. For any node $i \notin K$, let H_i^K be the union of all the connected subgraphs split by K which do not contain node i. Let $d(a, b)$ denote the distance between nodes a and b. Then, the observability index v_i will be bounded below by:

$$v_i \geq \frac{|H_i^K| + \sum_{j \in K} d(i, j)}{|K|}. \tag{17}$$

For a graph with multiple vertex-cut sets, the above statement must be satisfied for each of those sets. Before proving this statement, we first prove the following lemma:

Lemma 1. *The matrix created by concatenating any subset of columns in O_i must have full column-rank.*

Proof. To calculate $f = I_n$, O_i must have full column-rank, thus its columns are mutually independent. Therefore, any matrix created by concatenating any subset of columns of O_i must also have mutually independent columns, hence it has full column-rank.

For any subset of nodes S, denote P_S the matrix created by concatenating the columns of O_i corresponding to the nodes in S. P_S must have full column-rank which is $|S|$, thus it must have $|S|$ independent rows as well. Apply to the case of H_i^K, the only source of independent rows in $P_{H_i^K}$ that node i can get is from nodes in K, formally $x_j(t)$ where $j \in K$ and $t > 0$ ($x_j(0)$ is the initial value of node j, thus contains no information about H_i^K). Any observation at node i could be expressed by a linear combination of $x_j(t)$ for all $t > 0$ and $x_a(0)$ where $a \notin H_i^K$. Node i needs to observe $|H_i^K|$ independent rows in total, therefore it must run observations for at least $|H_i^K|/|K|$ iterations (this is the lower bound because not every single $x_j(t)$ value is independent, there may be a tighter bottleneck inside H_i^K). However, the value $x_j(t)$ must take at least $d(i,j)$ iterations to reach the node i, so in total, the lower bound for v_i is $\left(|H_i^K| + \sum_{j \in K} d(i,j) \right) / |K|$.

Example: Let $|H_i^K| = 5$, $K = \{a, b\}$, $d(i,a) = 3$, $d(i,b) = 4$. Then:

$$v_i \geq \frac{|H_i^K| + \sum_{j \in K} d(i,j)}{|K|} = \frac{5 + 3 + 4}{2} = 6.$$

This example is illustrated in Table 1. The only source of independent rows in H_i^K is from a and b, thus the node i must observe at least $|H_i^K| = 5$ values of $x_a(t)$ or $x_b(t)$ (with $t > 0$). The number of entries in columns $a + b$ (including blank entries) is equal to the numerator $|H_i^K| + \sum_{j \in K} d(i,j)$. To get the number of iterations, it is divided by the number of columns $|K|$.

Table 1. Illustration of bottleneck example

Iteration	a	b
0	$d(i,a)$	$d(i,b)$
1		
2		
3	$x_a(1)$	
4	$x_a(2)$	$x_b(1)$
5	$x_a(3)$	$x_b(2)$

Remark 1. Consider a graph with an articulation vertex (vertex-cut set with size 1), let q be that vertex, then the lower bound takes a very simple form:

$$v_i \geq |H_i^{\{q\}}| + d(i, q). \tag{18}$$

Remark 2. The quotient $(n - 1)/\deg_i$ is the special case of Eq. (17), where the vertex-cut set K is the set of neighbor nodes $\mathcal{N}_i$. In that case, $|H_i^K| = n - |\mathcal{N}_i| - 1$ and $d(i, j) = 1, \forall j \in K$. Thus, the bound is:

$$v_i \geq \frac{|H_i^K| + \sum_{j \in K} d(i, j)}{|K|} = \frac{n - |\mathcal{N}_i| - 1 + \sum_{j \in \mathcal{N}_i} 1}{|\mathcal{N}_i|} = \frac{n - 1}{\deg_i}.$$

Remark 3. This bound also covers the eccentricity bound (diameter bound) if K is allowed to be any subset of nodes excluding i (not necessarily a vertex cut set). Let K be any set which contains a single node j which is not i, $K = \{j\}$ may not split the graph, so $|H_i^{\{j\}}|$ might be 0. But the bound also has the distance term, so the observability index v_i must be at least $d(i, j)$. For all j in the graph excluding i, the bound becomes the eccentricity of the node i.

4.4 Exact Values for the Observation Phase and Observability Index

Currently, in the general case, the number of iterations v of the consensus phase or the observability index v_i can only be described formally in terms of upper and lower bounds. However the observability index v_i can be found empirically for all graphs in an $O(n^3)$ time-complexity algorithm, where n is the number of nodes in the graph. The Algorithm 2 below computes the observability index v_i of a given node i. To get the number of iterations of the observation phase, repeat the algorithm for each node i in the graph and set $v = \max\{v_i\}$.

Algorithm 2. Find the observability index v_i

$W \leftarrow$ weight matrix with random values in $[-1, 1]$
$O_i \leftarrow C_i$
$t \leftarrow 1$
while $\rho(O_i) < n$ **do**
$\quad O_i \leftarrow \begin{bmatrix} O_i \\ C_i W^t \end{bmatrix}$
$\quad t \leftarrow t + 1$
end while
$v_i \leftarrow t$
return v_i

Table 2 lists the exact number of iterations v of the observation phase for some well-known graphs and families of graphs. As indicated in the last column of Table 2, most of the exact number of iterations of those families of graphs are derived analytically from bounds described in the previous section, and could be validated by matching the lower bound (diameter bound, quotient $(n-1)/\deg_i$, or bottleneck bound) with the upper bound described in Eq. (16). For non-regular graphs, the exact number of iterations can be computed empirically using Algorithm 2.

Table 2. Consensus time of well-known graphs

Graph	Consensus time	Remark
Complete graph of n nodes $(n \geq 2)$	1	All nodes are pairwise neighbors
Path of n nodes $(n \geq 2)$	$n-1$	Diameter bound
Ring (cycle) of n nodes $(n \geq 3)$	$\lfloor n/2 \rfloor$	Diameter bound
Star of k leaves $(k \geq 3)$	k	Bottleneck at the center node
Wheel of k outer nodes $(k \geq 3)$	$\lceil k/3 \rceil$	$(n-1)/\deg_i$ for the outer nodes
Torus of $d \times d$ nodes $(d \geq 3)$	$\lceil (d^2-1)/4 \rceil$	$(n-1)/\deg_i$
Petersen graph	3	$(n-1)/\deg_i$
Möbius Kantor graph	5	$(n-1)/\deg_i$

The upper bound of Sundaram in Eq. (16) yields the exact value of the observability index in many cases. However in his PhD thesis [17] the author has conjectured (conjecture 1) that the bound is also a minimum and therefore exact bound. We provide a few counter examples which invalidate conjecture 1 in [17]. We found these counter examples by enumerating all the spanning trees decompositions in order to find that decomposition for which the largest branch is the smallest among all decompositions and comparing with the exact value from Algorithm 2. We found mismatches for the graphs in Fig. 2 where the upper bound based on the conjecture in [17] is not the exact value. The largest node in each of these graphs corresponds to node i whose observability index v_i is considered. The highlighted (thick) edges are the edges that belong to the spanning tree with the largest branch of minimum size. The upper bound would be the size of that largest branch. The caption x/exact for each graph shows x the value of the upper bound according to Eq. (16) and the exact value found by using Algorithm 2.

5   Implementations and Tests

In this section we briefly review the implementation of the network intrusion detection system based on an asymptotic consensus algorithm as proposed in

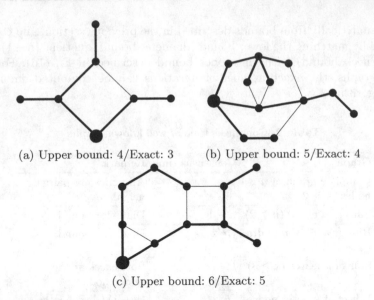

(a) Upper bound: 4/Exact: 3   (b) Upper bound: 5/Exact: 4

(c) Upper bound: 6/Exact: 5

Fig. 2. Plots of graphs where the upper bound is not the exact value

[18]. Next we describe the implementation of the finite time consensus algorithm in Sect. 3 for the same network intrusion detection system. Test data are described briefly and finally results and their analysis are provided. The APPENDIX provides a motivation for applying average consensus in the context of a distributed network intrusion detection system.

5.1 Asymptotic Consensus

In the consensus-based network intrusion detection system in [18], each node of the consensus network observes and analyzes the local traffic of a sub-network. The analysis is "anomaly based" using a naive Bayes classifier. The analysis yields likelihoods $P(O_i|h)$ that the locally observed traffic by node i falls under each of two hypotheses: traffic is anomalous h_a and traffic is normal h_n. The consensus problem consists to make the same determination for the whole computer network by averaging the sum of the local measurements. This determination is made by computing the joint likelihood $\prod_{i=1}^{n} P(O_i|h)$. The joint likelihood can be computed as a sum of log likelihoods $\sum_{i=1}^{n} \log P(O_i|h)$, allowing the average sum of the log likelihoods to be computed distributively by an average sum consensus algorithm. The phase where the joint likelihood $\prod_{i=1}^{n} P(O_i|h)$ is computed is called *consensus phase*. The consensus phase is in fact n parallel consensus loops, one for each node of the consensus network, where each consensus loop executes Eq. (1).

The initial values of the nodes in the consensus network are computed as $x_i(0) = \log P(O_i|h)$. The asymptotic consensus algorithm converges theoretically over an infinite number of iterations, at implementation level a *stopping condition* is added to the iterative scheme of Eq. (1). There are different stopping conditions possible, all approximate the closeness of the asymptotic consensus value to the real consensus value. We define the stopping condition as $|x_i(k+1) - x_i(k)| < \epsilon$, i.e. a consensus loop stops once the change in x_i from iteration k to iteration $k+1$ is smaller than a pre-defined threshold value ϵ. For weight matrices W satisfying convergence assumptions, the value $|x_i(k+1) - x_i(k)|$ decreases asymptotically as $k \to \infty$, so this difference always get eventually smaller than ϵ. In principle, the smaller ϵ is, the closer $x_i(k)$ is to the true average sum, the stronger the consensus is, but also the larger the number of iterations of the consensus loop is likely to be. Let c_i denotes the number of iterations executed by the consensus loop of node i prior to satisfy its stopping condition. Then the *number of iterations* ϕ of a consensus phase is given by

$$\phi = \max\{c_1, c_2, \ldots, c_n\}. \tag{19}$$

The consensus phase is synchronous, all nodes must have completed the consensus loop at iteration k before proceeding to execute the consensus loop iteration $k + 1$. A node i is assumed to have completed its consensus loop at iteration k if $k > c_i$.

Once a consensus phase is completed, each consensus node i has an approximation $\approx P(O|h_n)_i$ of the true likelihood $P(O|h_n) = \prod_{i=1}^{n} P(O_i|h_n)$, the likelihood the network wide observed network traffic is benign, and an approximation $\approx P(O|h_a)_i$ of the true likelihood $P(O|h_a) = \prod_{i=1}^{n} P(O_i|h_a)$, the likelihood the network wide observed network traffic is malicious. The asymptotic consensus program running at the level of node i outputs the ratio $\frac{\approx P(O|h_a)_i}{\approx P(O|h_n)_i}$, node i make a final decision to raise an alert based on $\frac{\approx P(O|h_a)_i}{\approx P(O|h_n)_i} > \tau$ where τ is some alert criterion.

5.2   Finite Time Consensus Implementation

The implementation of traffic reading and Bayesian analysis for the finite time consensus is the same as for the asymptotic consensus implementation. The Bayesian analysis of the local network traffic executed by node i returns two values: $P(O_i|h_a)$, the likelihood that the observed traffic at node i is anomalous; $P(O_i|h_n)$, the likelihood the observed traffic at node i is normal. As local analysis returns two measurements, each node i of the consensus network has two initial states: $x_i^A(0) = \log(P(O_i|h_a))$ and $x_i^N(0) = \log(P(O_i|h_n))$. The values x_i^N and x_i^A are respectively measurements of the local traffic observed by node i as normal and anomalous. The finite time consensus algorithm computes the consensus function for these two set of input measurements.

During the observation phase, observations for these two set of inputs are stored separately. Mathematically, at iteration k of the observation phase, each node i stores $C_i W^k x^A(0)$ in $y_i^A(k)$ and $C_i W^k x^N(0)$ in $y_i^N(k)$. The vector y_i^A records observations made by node i of the information flowing through it. At iteration $k = 0$, $y_i^A(0)$ is the likelihood that local traffic of nodes i and $j \in \mathcal{N}_i$ is anomalous. Similarly, y_i^N records observations made by node i of information flowing through it, $y_i^N(0)$ is the likelihood that local traffic of nodes i and $j \in \mathcal{N}_i$ is normal.

After the observation phase, the consensus values $P(O|h)_i$ for a given node i are computed either as $P(O|h_a)_i = \Gamma_i y_i^A$ and $P(O|h_n)_i = \Gamma_i y_i^N$ or from the inferred vectors $x^A(0)$ and $x^N(0)$. Similarly to asymptotic consensus, the finite time consensus procedure of node i returns the consensus ratio $\frac{P(O|h_a)_i}{P(O|h_n)_i}$. Each node i *decides* whether to raise an alert or not based on the consensus ratio $\frac{P(O|h_a)_i}{P(O|h_n)_i} > \tau$ for some predefined alert value ratio τ.

5.3   Simulation Environment

We have run tests for 11 different consensus network topologies: rings with 9, 25, 49, 81 and 121 sensors, 2-D torus with 9, 25, 49, 81 and 121 sensors and the Petersen graph (10 nodes 15 edges). Each run takes in input a test set and a graph representing a consensus network topology. Network traffic for the test set come from the NLS-KDD data set [34], an improved version of KDD'99 data set. The KDD'99 data set has been generated by the MIT Lincoln Laboratory for the evaluation of computer network intrusion detection systems under the sponsorships of the Defense Advanced Research projects Agency (DARPA) and the Air force Research Laboratory (AFRL) [35,36]. As for the KDD'99 data set, traffic representation and analysis from the NLS-KDD data set is based on 41 features. Note, we have filtered attacks in the NSL-KDD data set to retain only denial of service attacks.

A test consists for each node of a network topology to make 1000 network traffic readings, each reading come from data recorded in the NLS-KDD data set. After each local network traffic reading, an Bayesian analysis is performed, then, for asymptotic consensus, the consensus loop is run by each node until it satisfies its stopping condition, and for finite time consensus, the observation and inference phases are run independently for each node. For finite time consensus, prior to running a test, the matrices O_i, C_i, Γ_i are pre-computed and each matrix Γ_i is sent to the corresponding node i.

5.4   Test Results and Interpretations

Table 3 compares the convergence speed of the asymptotic and finite time consensus procedures. This table has 11 rows corresponding to the 11 consensus network topologies that have been tested. For asymptotic consensus, for each network topology, 1000 consensus phases have been run, each with a different initial vector $x(0)$. For each consensus phase, we have recorded the convergence speed, i.e. the number ϕ of iterations during the consensus phase. Table 3 reported the average number of iterations (rounded) for each consensus network topology.

For the finite time consensus, the tests conducted in this section are based on the two versions of the finite time consensus: 1- the consensus function is computed directly from the vector y_i; 2- $x(0)$ is inferred from y_i, then the consensus function is computed. For the version 1, we have run tests with two set of values defining the length of observation phases: one based on the upper bound defined in Eq. (15) and a second one based on lower bounds defined by the diameter of the network topology. For version 2, we use the exact bounds provided in Table 2. The results in Table 3 for the finite time consensus come from those bounds. Nonetheless, we have run tests to confirm that the consensus problem is solved inside the number of iterations listed in Table 3. Each test consists to execute the finite time algorithm 1000 times, this for each consensus network topology.

In Table 3, the test results are in the columns "Asymptotic", "Exact-v1" and "Exact-v2" corresponding respectively to tests with the asymptotic consensus algorithm, the finite time consensus algorithm computing the consensus function directly from y_i and finally the finite time consensus algorithm computing the consensus function from the inferred $x(0)$ vector. In column "Exact-v1", a/b represents respectively the lower bound and the upper bound on the length of the sequence of observations allowed to compute the consensus values.

Table 3. The number of iterations to compute consensus

Topology	Size	Asymptotic	Exact-v1	Exact-v2
Ring	9	25	5/7	5
	25	146	13/23	13
	49	452	25/47	25
	81	990	41/79	41
	121	1772	61/119	61
Torus	9	7	3/7	2
	25	16	5/23	6
	49	28	7/47	12
	81	43	9/79	20
	121	59	11/119	30
Petersen	10	9	2/8	3

One obvious observation from the column Exact-v1 in Table 3 is that upper bounds based on Eq. (15) are very poor approximations of the number of steps executed during the observation phases as each network topology computes consensus inside the corresponding lower bound. Considering that lower bounds in the column Exact-v1 are tighter approximations, we conclude from Table 3 and ring networks that the length of the observation sequence of both versions of the finite time consensus is considerably smaller than the number of iterations of the consensus phases of asymptotic consensus. For 2-D torus, the gains of the finite time consensus are more moderated. The larger connectivity of 2-D torus compared to ring networks benefits more greatly to asymptotic consensus than it does to the finite time consensus. We note in Table 3 that version v2, which first infers $x(0)$ from y_i, needs more observations. This is generally the case, solving for the identity matrix requires a larger observability matrix (row wise) compared to solving for the matrix form of the consensus function. Nonetheless, these results show that implementing the consensus-based intrusion detection system using any of the two version of the finite time consensus reduces substantially the time needs to compute consensus for these network topologies.

Computing the consensus function from a vector $x(0)$ inferred from observed data is computationally more expensive. The benefit is that any or many consensus functions can be computed from the same observation phase. Further, the results in the column "Exact-v2" of Table 3 are the same no matter which consensus function is computed. The length of the observation phase depends only on the consensus network topology and the number of nodes in the network.

It should be noted that the above good performances of the finite time consensus are limited to static networks. The topology of the consensus networks may change following the physical failure of a link or a node or following cyber attacks on the network intrusion detection infrastructure. For example, if the link between nodes 1 and 2 in Fig. 1 fails, the degree of these two nodes is reduced, consequently matrices C_1 and C_2 have to be re-computed as well as the observability and the Γ matrices for these two nodes. These re-computations will require time, but they can be done locally. However, for bounds that are based on the graph diameter or the max degree of the nodes in the consensus network, these cannot be updated locally, so the number of iterations in the following consensus observation phases will be set by upper bounds only. A second point to raise, improvements in communication to compute cnensus come at the cost of an increase in terms of the consensus node storage capabilities and one matrix multiplication during the inference phase.

6   Conclusion

In this paper we have introduced new results improving the convergence speed of finite time consensus algorithms derived from the observability matrix of the consensus network. We have also introduced invariant properties of the observability matrix which directly extend into new algorithmic variants of this finite time consensus algorithm. These properties may as well support design strategies to render these consensus algorithms more robust against data falsification and Byzantine attacks. We have implemented a network intrusion detection system using the finite time consensus and compared its performance with the same network intrusion detection system based on an asymptotic average consensus algorithm. Tests show that the number of steps needed to obtain a finite time consensus is smaller compared to an acceptable approximation of consensus with an asymptotic consensus algorithm. Tests also validate analytical results about the behavior of exact consensus algorithms.

The intrusion detection system describes in [18] and the one in this paper are computer network security systems based on the cooperation of several devices or sub-systems. Considering the diversify of computer network attacks [37], such cooperative systems have to aggregate relevant information for a subset as large as possible of those threats. Finite time consensus algorithms based on the observability matrix can compute consensus for any and many consensus functions from the same observation phase. It is certainly worth to investigate whether this property can be applied to design far more versatile cooperative intrusion detection systems.

Several proposals have been put forward in the control theory community to improve the robustness of consensus algorithms based the observability matrix against network instabilities caused by hardware failures, cyber attacks or simply by spontaneous reconfigurations in some wireless networks. All these proposals are combinatorial in nature, only one or two network changes can be handled effectively, otherwise the proposed solutions are too computationally expensive. Improving security and adapting these finite time consensus algorithms to dynamic networks is a second focus of our future work.

Acknowledgments. Funding for this project comes from the Professorship Start-Up Support Grant VGU-PSSG-02 of the Vietnamese-German University. The authors thank this institution for supporting this research.

Appendix

Network intrusion detection systems monitor computer network infrastructures, seeking to identify malicious intends through the analysis of network traffic. Typically detection is broken down into two phases: an observation phase where network traffic information is collected and an analysis phase where the observed traffic is analyzed and categorized into benign or malignant network traffic. Traffic observation is performed through sensors that collect information about specific features of network traffic. As today's computer networks are quite large, composed of several heterogeneous sub-networks, traffic observation often needs to be done distributively with sensors placed at different strategic locations. Figure 3 exemplifies this situation, showing sensors distributed across a supervised network, where each sensor is in charge of observing the local traffic of a sub-network.

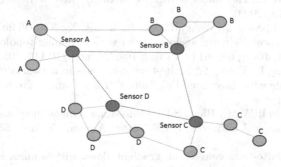

Fig. 3. Topology of a network intrusion detection system.

Typically, in a large computer network, sensors are doing more than simply observing local traffic in sub-networks, they also perform analysis of the local sub-network traffic. In that case, sensors are full scale sensing and analytical devices. As traffic analysis is performed at the level of each sensor, local traffic can be classified and remedial actions can be taken if an intrusive behavior is detected. However, in some cases, a certain degree of aggregation of local analysis results can be helpful to address for example attacks from concurrent sources such as distributed denial of service (DDoS), to develop network wide coordinated responses to attacks or simply to increase the detection accuracy of each local analysis based on information from other sub-networks. If data aggregation is a key component of an intrusion detection system then this component must be designed to maintain the survivability and robustness of the system. Consensus algorithms are a relevant choice in this context as they provide protocols to compute aggregation functions in a completely distributed manner, eliminating issues such as single point of failure and others related to centralized computing of network data.

Toulouse et al. [18] have introduced a network intrusion detection system in which aggregation and fusion of local traffic analysis is computed distributively using an average consensus algorithm. In order to accomplish this task, sensors communicate with each other through what we call a *consensus network*. In Fig. 3, the consensus network is a ring network linking the four sensor nodes. While exchanging information with their neighbors in the consensus network, nodes repeatedly average a sum of values representing their diagnostic about the state of the local traffic as well as the diagnostic of their neighbors. Through local averaging, sensors approximate a *consensus value* which is used as measurement of some relevant network wide state of the monitored network system.

References

1. Xiao, L., Boyd, S., Kim, S.-J.: Distributed average consensus with least-mean-square deviation. J. Parallel Distrib. Comput. **67**(1), 33–46 (2007)
2. Xiao, L., Boyd, S., Lall, S.: A scheme for robust distributed sensor fusion based on average consensus. In: Fourth International Symposium on Information Processing in Sensor Networks, IPSN 2005, April 2005, pp. 63–70 (2005)
3. Xiao, L., Boyd, S.: Fast linear iterations for distributed averaging. Syst. Control. Lett. **53**, 65–78 (2003)
4. Wang, X., Li, J., Xing, J., Wang, R., Xie, L., Zhang, X.: A novel finite-time average consensus protocol for multi-agent systems with switching topology. Trans. Inst. Meas. Control. **40**(2), 606–614 (2018). https://doi.org/10.1177/0142331216663617
5. Xiao, F., Wang, L., Chen, T.: Finite-time consensus in networks of integrator-like dynamic agents with directional link failure. IEEE Trans. Autom. Control. **59**(3), 756–762 (2014)
6. Hui, Q., Haddad, W.M., Bhat, S.P.: Finite-time semistability and consensus for nonlinear dynamical networks. IEEE Trans. Autom. Control. **53**(8), 1887–1900 (2008)
7. Cortés, J.: Finite-time convergent gradient flows with applications to network consensus. Automatica **42**(11), 1993–2000 (2006). http://www.sciencedirect.com/science/article/pii/S000510980600269X

8. Ko, C.-K., Gao, X.: On matrix factorization and finite-time average-consensus. In: Proceedings of the 48th IEEE Conference on Decision and Control (CDC) Held Jointly with 2009 28th Chinese Control Conference, pp. 5798–5803, December 2009

9. Ko, C.-K.: On matrix factorization and scheduling for finite-time average-consensus. Ph.D. dissertation, California Institute of Technology (2010)

10. Kibangou, A.Y.: Graph Laplacian based matrix design for finite-time distributed average consensus. In: 2012 American Control Conference (ACC), pp. 1901–1906, June 2012

11. Dung, T.M., Alain, T., Kibangou, Y.: Distributed design of finite-time average consensus protocols. In: IFAC Proceedings Volumes 2013 4th IFAC Workshop on Distributed Estimation and Control in Networked Systems, vol. 46, no. 27, pp. 227–233 (2013). http://www.sciencedirect.com/science/article/pii/S1474667015402320

12. Sundaram, S., Hadjicostis, C.N.: Finite-time distributed consensus in graphs with time-invariant topologies. In: 2007 American Control Conference, pp. 711–716, July 2007

13. Yuan, Y., Stan, G.-B., Shi, L., Barahona, M., Goncalves, J.: Decentralised minimum-time consensus. Automatica 49(5), 1227–1235 (2013)

14. Charalambous, T., Yuan, Y., Yang, T., Pan, W., Hadjicostis, C.N., Johansson, M.: Decentralised minimum-time average consensus in digraphs. In: 52nd IEEE Conference on Decision and Control, pp. 2617–2622, December 2013

15. Sundaram, S., Hadjicostis, C.N.: Distributed consensus and linear functional calculation in networks: an observability perspective. In: 2007 6th International Symposium on Information Processing in Sensor Networks, pp. 99–108, April 2007

16. Sundaram, S., Hadjicostis, C.N.: Distributed function calculation and consensus using linear iterative strategies. IEEE J. Sel. Areas Commun. 26(4), 650–660 (2008). https://doi.org/10.1109/JSAC.2008.080507

17. Sundaram, S.: Linear Iterative Strategies for Information Dissemination and Processing in Distributed Systems. University of Illinois at Urbana-Champaign (2009). https://books.google.com.vn/books?id=K-LOAQAACAAJ

18. Toulouse, M., Minh, B.Q., Curtis, P.: A consensus based network intrusion detection system. In: 2015 5th International Conference on IT Convergence and Security (ICITCS), pp. 1–6, August 2015

19. Bauso, D., Giarré, L., Pesenti, R.: Distributed consensus in noncooperative inventory games. Eur. J. Oper. Res. 192(3), 866–878 (2009). http://www.sciencedirect.com/science/article/pii/S0377221707010181

20. Akyildiz, I.F., Lo, B.F., Balakrishnan, R.: Cooperative spectrum sensing in cognitive radio networks: a survey. Phys. Commun. 4(1), 40–62 (2011). https://doi.org/10.1016/j.phycom.2010.12.003

21. Li, Z., Yu, F.R., Huang, M.: A cooperative spectrum sensing consensus scheme in cognitive radios. In: IEEE INFOCOM 2009, pp. 2546–2550, April 2009

22. Olfati-Saber, R., Murray, R.M.: Consensus problems in networks of agents with switching topology and time-delays. IEEE Trans. Autom. Control. 49(9), 1520–1533 (2004). https://doi.org/10.1109/tac.2004.834113

23. Chen, J., Patton, J.R., Zhang, H.-Y.: Design of unknown input observers and robust fault detection filters. Int. J. Control. 63(1), 85–105 (1996)

24. Isermann, R.: Model-based fault-detection and diagnosis - status and applications. Annu. Rev. Control. 29, 71–85 (2005)

25. Teixeira, A., Sandberg, H., Johansson, K.H.: Networked control systems under cyber attacks with applications to power networks. In: Proceedings of the 2010 American Control Conference, pp. 3690–3696, June 2010

26. Pasqualetti, F., Bicchi, A., Bullo, F.: Consensus computation in unreliable networks: a system theoretic approach. IEEE Trans. Autom. Control. **57**(1), 90–104 (2012)
27. Pasqualetti, F., Dörfler, F., Bullo, F.: Attack detection and identification in cyberphysical systems. IEEE Trans. Autom. Control. **58**(11), 2715–2729 (2013)
28. Pasqualetti, F., Bicchi, A., Bullo, F.: On the security of linear consensus networks. In: Proceedings of the 48th IEEE Conference on Decision and Control (CDC) Held Jointly with 2009 28th Chinese Control Conference, pp. 4894–4901, December 2009
29. Silvestre, D., Rosa, P., Cunha, R., Hespanha, J.P., Silvestre, C.: Gossip average consensus in a byzantine environment using stochastic set-valued observers. In: 52nd IEEE Conference on Decision and Control, pp. 4373–4378, December 2013
30. Pasqualetti, F., Bicchi, A., Bullo, F.: Distributed intrusion detection for secure consensus computations. In: 2007 46th IEEE Conference on Decision and Control, pp. 5594–5599, December 2007
31. Ahlswede, R., Cai, N., Li, S.Y., Yeung, R.W.: Network information flow. IEEE Trans. Inf. Theor. **46**(4), 1204–1216 (2000)
32. Ghosh, S., Lee, J.: Optimal synthesis for finite-time consensus under fixed graphs. In: 2011 50th IEEE Conference on Decision and Control and European Control Conference, pp. 2052–2057, December 2011
33. Ghosh, S., Lee, J.W.: Optimal distributed consensus on unknown undirected graphs. In: 2012 IEEE 51st IEEE Conference on Decision and Control (CDC), pp. 2244–2249, December 2012
34. Tavallaee, M., Bagheri, E., Lu, W., Ghorbani, A.A.: A detailed analysis of the KDD cup 99 data set. In: 2009 IEEE Symposium on Computational Intelligence for Security and Defense Applications, pp. 1–6, July 2009
35. Lippmann, R., Haines, J.W., Fried, D.J., Korba, J., Das, K.: Analysis and results of the 1999 DARPA off-line intrusion detection evaluation. In: Debar, H., Mé, L., Wu, S.F. (eds.) RAID 2000. LNCS, vol. 1907, pp. 162–182. Springer, Heidelberg (2000). https://doi.org/10.1007/3-540-39945-3_11. http://dl.acm.org/citation.cfm?id=645838.670722
36. Lippmann, R., et al.: Evaluating intrusion detection systems: the 1998 DARPA off-line intrusion detection evaluation. In: Proceedings of DARPA Information Survivability Conference and Exposition, DISCEX 2000, pp. 12–26 (2000)
37. Kotenko, I., Doynikova, E.: Selection of countermeasures against network attacks based on dynamical calculation of security metrics. J. Def. Model. Simul. **15**(2), 181–204 (2018). https://doi.org/10.1177/1548512917690278

Parallel Learning Algorithms of Local Support Vector Regression for Dealing with Large Datasets

Thanh-Nghi Do[1,2(✉)] and Le-Diem Bui[1,3]

[1] College of Information Technology, Can Tho University, Cantho 92100, Vietnam
dtnghi@cit.ctu.edu.vn
[2] UMI UMMISCO 209 (IRD/UPMC), UPMC, Sorbonne University,
Pierre and Marie Curie University, Paris 6, France
[3] AI Lab, Computer Science Department, Gyeongsang National University, Jinju, Korea

Abstract. New parallel algorithms of local support vector regression (local SVR), called kSVR, krSVR are proposed in this paper to efficiently handle the prediction task for large datasets. The learning strategy of kSVR performs the regression task with two main steps. The first one is to partition the training data into k clusters, followed which the second one is to learn the SVR model from each cluster to predict the data locally in the parallel way on multi-core computers. The krSVR learning algorithm trains an ensemble of T random kSVR models for improving the generalization capacity of the kSVR alone. The performance analysis in terms of the algorithmic complexity and the generalization capacity illustrates that our kSVR and krSVR algorithms are faster than the standard SVR for the non-linear regression on large datasets while maintaining the high correctness in the prediction. The numerical test results on five large datasets from UCI repository showed that proposed kSVR and krSVR algorithms are efficient compared to the standard SVR. Typically, the average training time of kSVR and krSVR are 183.5 and 43.3 times faster than the standard SVR; kSVR and krSVR also improve 62.10%, 63.70% of the relative prediction correctness compared to the standard SVR, respectively.

Keywords: Support vector regression (SVR) ·
Local support vector regression (local SVR) · Ensemble learning ·
Large datasets

1 Introduction

In last decades, the progress in computer hardware, the increasing number of internet users and mobile device access to internet have enabled explosion in data. Researchers from the University of Berkeley estimate that about 1 Exabyte (10^9 Gigabyte) of data are generated every year [1]. Recent book [2] shows that Google, Yahoo!, Microsoft, Facebook, Twitter, YouTube and other internet-based companies have Exabytes of data due to hundreds of millions of users and billion daily active users. Such a huge amount of data yields challenges in data analysis because current analytical techniques are not

© Springer-Verlag GmbH Germany, part of Springer Nature 2019
A. Hameurlain et al. (Eds): TLDKS XLI, LNCS 11390, pp. 59–77, 2019.
https://doi.org/10.1007/978-3-662-58808-6_3

well suited for that data scale. Therefore, it is a high priority to create new learning algorithms for addressing massive datasets. Our research aims to propose parallel learning algorithms of local support vector regression (local SVR) for dealing with large datasets.

Support vector machines (SVM) proposed by [3] and kernel-based methods have shown practical relevance for classification, regression and novelty detection. Successful applications are reported for face identification, text categorization and bioinformatics [4]. Nevertheless, the SVM learning is accomplished through a quadratic programming (QP), so that the computational cost of a SVM approach is at least square of the number of training datapoints making SVM impractical to handle large datasets. There is a need to scale-up SVM learning algorithms to deal with massive datasets.

In this paper, our investigation aims at developing new parallel algorithms of local SVR to efficiently handle the non-linear regression of large datasets.

Instead of building a global SVR model, as done by the classical algorithm is very difficult to deal with large datasets, our kSVR algorithm [5] is to learn in the parallel way an ensemble of local ones that are easily trained by the standard SVR algorithm like LibSVM [6]. The kSVR algorithm performs the training task with two main steps. The first one is to use $k$means algorithm [7] to partition the large training dataset into k clusters (subsets). The idea is to reduce the data size for training local non-linear SVR models at the second step. The algorithm learns k non-linear SVR models in the parallel way on multi-core computers in which a SVR model is trained in each cluster to predict the data locally. We propose to develop the new ensemble-based learning algorithm, called krSVR. The training task of krSVR learns the multiple random kSVR models to improve the generalization capacity of the kSVR alone.

The performance analysis in terms of the algorithmic complexity and the generalization capacity and the numerical test results on five large datasets from UCI repository [8] showed that our proposed kSVR and krSVR algorithms are faster than the standard SVR for the non-linear regression of large datasets while achieving the high correctness in the prediction.

The paper is organized as follows. Section 2 briefly introduces the SVR algorithm. Section 3 presents our proposed parallel algorithm kSVR of local SVR models for the non-linear regression on large datasets. The learning algorithm krSVR for the ensemble of T random kSVR models is illustrated in Sect. 4. The experimental results is presented in Sect. 5 before the discussion on related works in Sect. 6. We then conclude in Sect. 7.

2 Support Vector Regression

Let's start with the regression task for a training dataset with m datapoints x_i ($i = 1, \ldots, m$) in the n-dimensional input space R^n, having corresponding targets $y_i \in R$, support vector regression (SVR) proposed by [3] tries to find the best hyperplane (denoted by the normal vector $w \in R^n$ and the scalar $b \in R$) that has at most ε deviation from the target value y_i. The SVR pursues this goal with the quadratic programming (1) (Fig. 1).

$$\min (1/2) \sum_{i=1}^{m} \sum_{j=1}^{m} (\alpha_i - \alpha_i^*)(\alpha_j - \alpha_j^*)K\langle x_i, x_j \rangle - \sum_{i=1}^{m} (\alpha_i - \alpha_i^*)y_i + \varepsilon \sum_{i=1}^{m} (\alpha_i + \alpha_i^*)$$

$$s.t. \begin{cases} \sum_{i=1}^{m} (\alpha_i - \alpha_i^*) = 0 \\ 0 \le \alpha_i, \alpha_i^* \le C \quad \forall i = 1, 2, ..., m \end{cases} \tag{1}$$

where C is a positive constant used to tune the margin and the error and a linear kernel function $K\langle x_i, x_j \rangle = \langle x_i \cdot x_j \rangle$

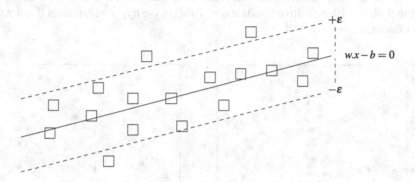

Fig. 1. Linear support vector regression

The support vectors (for which $\alpha_i, \alpha_i^* > 0$) are given by the solution of the quadratic programming (1), and then, the predictive hyperplane and the scalar b are determined by these support vectors. The prediction of a new datapoint x based on the SVR model is as follows:

$$predict(x, SVR\ model) = \sum_{i=1}^{\#SV} (\alpha_i - \alpha_i^*)K\langle x, x_i \rangle - b. \tag{2}$$

SVR algorithms can use different kernel functions [9] to handle variations on prediction problems. It only needs replacing the usual linear kernel function $K\langle x_i, x_j \rangle = \langle x_i \cdot x_j \rangle$ with other popular non-linear kernel functions, including:

- a polynomial function of degree d: $K\langle x_i, x_j \rangle = (\langle x_i \cdot x_j \rangle + 1)^d$.
- a RBF (Radial Basis Function): $K\langle x_i, x_j \rangle = e^{-\gamma \|x_i - x_j\|^2}$.

The SVR models are most accurate and practical relevance for many successful applications reported in [4].

3   Parallel Algorithm of Local Support Vector Regression

The study in [10] illustrated that the computational cost requirements of the SVR solutions in (1) are at least $O(m^2)$ (where m is the number of training datapoints), making standard SVM intractable for large datasets. Learning a global SVR model on the full massive dataset is challenge due to the very high computational cost.

3.1   Learning k Local SVR Models

Our proposed kSVR algorithm learns an ensemble of local SVR models that are easily trained by the standard SVR algorithm. As illustrated in Fig. 3, the kSVR handles the regression task with two main steps. The first one uses $k$means algorithm [7] to partition the full training set into k clusters, and then the second one trains an ensemble of local SVR models in which a non-linear SVR is learnt from each cluster to predict the data locally. We consider a simplest regression task given a target variable y and a predictor (variable) x. Figure 2 shows the comparison between a global SVR model (left part) and 3 local SVR models (right part) for this regression task, using a non-linear RBF kernel function with $\gamma = 10$, a positive constant $C = 10^5$ (i.e. the hyper-parameters $\theta = \{\gamma, C\}$) and a tolerance $\varepsilon = 0.05$.

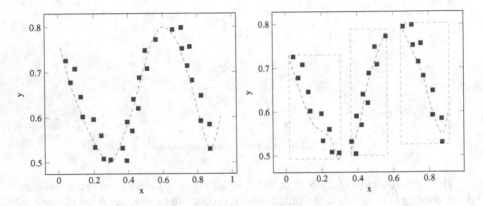

Fig. 2. Global SVR model (left part) versus 3 local SVR models (right part)

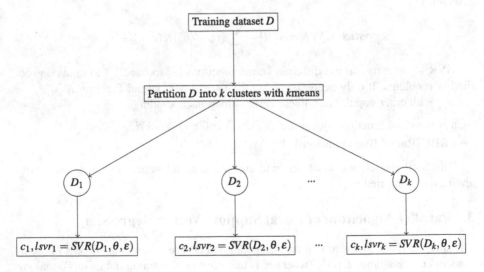

Fig. 3. Training k local SVR models (kSVR)

Algorithm 1. Parallel training algorithm $kSVR(D, k, \gamma, C, \varepsilon)$ for k local SVR

---

    input :

        training dataset D
        number of local models k
        tolerance ε
        hyper-parameter γ of RBF kernel function
        positive constant C for tuning margin and errors

    output:

        k local SVR models

1 **begin**
2   /*$k$means performs in the parallel way (only E-step can be directly parallelized)
     the data clustering on dataset D;*/
3   creating k clusters denoted by $D_1, D_2, \ldots, D_k$ and
4   their corresponding centers $c_1, c_2, \ldots, c_k$
5   #pragma omp parallel for schedule(dynamic)
6   **for** $i \leftarrow 1$ **to** k **do**
7 | /*learning local support vector regression model from D_i;*/
8 | $lsvr_i = svr(D_i, \gamma, C, \varepsilon)$
9   **end**
10   return $kSVR$-model $= \{(c_1, lsvr_1), (c_2, lsvr_2), \ldots, (c_k, lsvr_k)\}$
11 **end**

Since the cluster size is smaller than the full training data size, the standard SVR can easily perform the training task on the data cluster. Furthermore, the $kSVR$ learns independently k local models from k clusters. This training task of $kSVR$ is easily parallelized to take into account the benefits of high performance computing, e.g. multi-core computers or grids. The simplest development of the parallel $kSVR$ algorithm is based on the shared memory multiprocessing programming model OpenMP [11] on multi-core computers. The parallel training of $kSVR$ is described in Algorithm 1.

3.2 Prediction of a New Datapoint x Using k Local SVR Models

The $kSVR$-model $= \{(c_1, lsvr_1), (c_2, lsvr_2), \ldots, (c_k, lsvr_k)\}$ is used to predict the target value of a new datapoint x as follows. The first step is to find the closest cluster based on the distance between x and the cluster centers:

$$c_{NN} = \arg\min_c \; distance(x, c). \tag{3}$$

And then, the target value of x is predicted by the local SVR model $lsvr_{NN}$ (corresponding to c_{NN}):

$$predict(x, kSVR \; model) = predict(x, lsvr_{NN}). \tag{4}$$

We can see that the prediction of a new datapoint x in k local SVR models depends not only on the clustering assignment but also on the prediction of the local SVR model at the final stage. It is the local SVR model that directly predicts the outcome of the new datapoint x.

We study the performance analysis in terms of the algorithmic complexity and the generalization capacity to justify the competence of the learning algorithm kSVR.

3.3 Performance Analysis

Let's start analysing the algorithmic complexity of building k local SVR models with the parallel kSVR algorithm. The full dataset with m datapoints is partitioned into k balanced clusters. It means that the cluster size is about $\frac{m}{k}$. According to [10], the computational cost requirements of the SVR solutions is quadratic in the number of training datapoints. Therefore, the training complexity of a local SVR is $O(\frac{m^2}{k^2})$. Thus, the algorithmic complexity of parallel training k local SVR models on a P-core processor is:

$$O(\frac{k}{P}\frac{m^2}{k^2}) = O(\frac{m^2}{k \cdot P}).\tag{5}$$

This complexity analysis illustrates that parallel learning k local SVR models in the kSVR algorithm[1] is $k.P$ times faster than building a global SVR model (the complexity is at least $O(m^2)$).

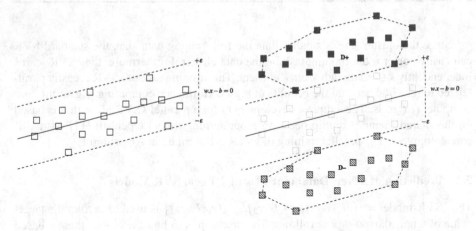

Fig. 4. The regression task of SVR (left part) can be regarded as a binary classification problem of SVC (right part)

The generalization capacity of kSVR models trained by the kSVR algorithm can be explained in terms of the margin size of the binary support vector classification (SVC).

---

[1] It must be noted that the complexity of the kSVR approach does not include the $k$means clustering used to partition the full dataset. But this step requires insignificant time compared with the quadratic programming solution.

An intuitive geometric approach in [12] illustrates that the regression task of the SVR is considered as a classification problem of the SVC. Let x be the predictor variable and y the target variable. In the left part of Fig. 4, the SVR approach tries to find the optimal plane ($w.x - b = 0$) that has at most ε deviation from target values y_i. This problem is transformed into the binary classification one as illustrated in the right part of Fig. 4. The positive class $D+$ is formed by increasing the target values y_i by ε. The negative class $D-$ is formed by decreasing the target values y_i by ε. The optimal plane found by the SVC approach for separating $D+$ from $D-$ is identical to the solution of the SVR approach. The biggest margin solution (largest separation boundary of two classes) gives the safest prediction model. It means that we can explain the generalization capacity of kSVR models trained by the kSVR algorithm in terms of the largest margin size of the binary SVC.

The performance analysis in terms of the generalization capacity of such local models is illustrated in [13–15]. Recently Do and his colleague [16,17] show that the learning algorithms of local SVC models give a guarantee of the generalization capacity compared to global SVC one.

To assess the generalization capacity of local SVR models, we start with Theorem 5.2 ([18] p. 139). It mentions that the generalization ability of the large margin hyperplane is high. Given a training set with m datapoints being separated by the maximal margin hyperplanes, the expectation of the probability of test error is bounded as follows:

$$EP_{error} \leq E \left\{ min \left(\frac{sv}{m}, \frac{1}{m} \left[\frac{R^2}{\Delta} \right], \frac{n}{m} \right) \right\}. \tag{6}$$

where sv is the number of support vectors, R is the radius of the sphere containing the data and Δ is the size of the margin, n is the number of dimensions.

It means that the good generalization ability of the maximal margin hyperplane is illustrated in:

$$min \left(\frac{1}{m} \left[\frac{R^2}{\Delta} \right] \right). \tag{7}$$

In the training task, the kSVR algorithm splits the full dataset having m datapoints into k clusters (the cluster size m_k is about $\frac{m}{k}$). And then the generalization capacity of local SVR models is assessed in:

$$min \left(\frac{1}{m_k} \left[\frac{R_k^2}{\Delta_k} \right] \right). \tag{8}$$

The generalization analysis bases on the comparison between Eq. (7) of the global SVR model trained from the full dataset and Eq. (8) of the local SVR model learnt from a cluster (subset). According to Theorem 1 in [16] and Theorems 2, 3 in [17], the inequality $\Delta_{X_k} \geq \Delta_X$ holds. Nevertheless, the use of the subset in the training task of kSVR leads to $R_k \leq R$ and $m_k \leq m$. It illustrates that there exists a compromise between the locality (the subset size, the radius of the sphere containing the data) and the generalized capacity (the margin size). Therefore, local SVR models trained by the

kSVR algorithm can guarantee the prediction performance compared to the global SVR one.

The performance analysis in terms of the algorithmic complexity (5) and the generalization capacity (8) shows that the parameter k is used in the kSVR to give a trade-off between the generalization capacity and the computational cost. This can be understood as follows:

- If k is large then the kSVR algorithm reduces significant training time. And then, the size of a cluster is small; The locality is extremely with a very low capacity.
- If k is small then the kSVR algorithm reduces insignificant training time. However, the size of a cluster is large; It improves the capacity.

It leads to set k so that the cluster size is a large enough (e.g. 200 proposed by [14]).

4   Parallel Algorithm of T Random Local Support Vector Regression Models

The performance analysis of the kSVR algorithm shows that there is a trade-off between the algorithmic complexity and the generalization capacity. If the kSVR tries to speed up the training time against the learning algorithm of global SVR models by increasing the number of clusters (the parameter k) then it reduces the generalization ability. Due to

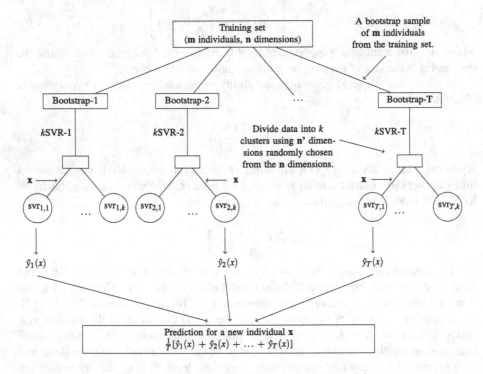

Fig. 5. Training algorithm of T random kSVR models

Algorithm 2. Parallel training algorithm of T random kSVR models

---

    input :

                training dataset D having m individuals in n dimensions

                number of kSVR models T

                k local models in kSVR

                n' random dimensions used for clustering in kSVR

                tolerance ε for a SVR model

                hyper-parameter γ of RBF kernel function

                positive constant C for tuning margin and errors

    output :

                T random kSVR models

  1 **begin**

  2   #pragma omp parallel for schedule(dynamic)

  3   **for** $t \leftarrow 1$ **to** T **do**

  4   Sampling a bootstrap D_t from D using n' dimensions randomly chosen

  5   from n original dimensions

  6   /*learning k local SVR models from D_t;*/

  7   kSVR$_t$ = kSVR$(D_t, k, \gamma, C, \varepsilon)$

  8   **end**

  9   return T random kSVR models = $\{k\text{SVR-}1, k\text{SVR-}2, \ldots, k\text{SVR-}T\}$

 10 **end**

---

this problem, we propose the ensemble-based learning algorithm of random local SVR models to improve the generalization capacity of the local one. The main idea is based on the Bias-Variance framework proposed by Breiman [19,20]. The performance of a learning model depends on Bias term and Variance term. Bias is the systematic error and not depending on the learning sample. Variance is the error with respect to the variability of the learning model due to the learning sample randomness. The main idea of ensemble-based learning algorithms [19–21] and [22] is to use the randomization of the learning sample for reducing Variance and/or Bias. This key idea leads to the improvement of the generalization capacity of the use of the single one model. Therefore, we propose the ensemble learning algorithm krSVR to train T random kSVR models for reducing Variance. It means that the krSVR improves the generalization ability of the kSVR model alone.

4.1 Learning T Random Local SVR Models

The krSVR algorithm trains the ensemble of T random kSVR models using the kSVR algorithm (described in Algorithm 1 and Fig. 3). The kSVR algorithm trains the tth kSVR model from the tth bootstrap sample (sampling with replacement from the original dataset) using n' dimensions randomly sampling without replacement from n original dimensions.

As described in Algorithm 2 and Fig. 5, the krSVR constructs independently T random local SVR models. It allows parallelizing the learning task with OpenMP on multi-core computers.

Thus, the complexity of parallel learning krSVR on a P-core processor is as follows:

$$O(\frac{T.m^2}{k.P}). \tag{9}$$

4.2 Prediction of a New Datapoint x Using T Random Local SVR Models

The prediction for a new individual x is the average of the prediction results obtained by T random kSVR models.

5 Evaluation

We are interested in the performance of new parallel algorithms of local SVR (kSVR, krSVR) for dealing with large datasets. There is a need of numerical test results in terms of training time and prediction correctness to be consistent with the performance analysis of the algorithmic complexity and the generalization capacity in Sect. 3.

5.1 Software Programs

We have implemented algorithms kSVR, krSVR in C/C++, OpenMP [11], using the Automatically Tuned Linear Algebra Software (ATLAS [23]) and the highly efficient standard library SVM, LibSVM [6].

Due to the consistency with the performance analysis of the algorithmic complexity and the generalization capacity, our evaluation is reported in terms of training time and prediction correctness. We are interested in the comparison the regression results obtained by our proposed kSVR, krSVR for local SVR models with LibSVM for global SVR models.

All experiments are run on machine Linux Fedora 20, Intel(R) Core i7-4790 CPU, 3.6 GHz, 4 cores and 32 GB main memory.

5.2 Datasets

All experiments are conducted with the five datasets from UCI repository [8]. Table 1 presents the description of datasets. The evaluation protocols are illustrated in the last column of Table 1. Datasets are already divided in training set (Trn) and test set (Tst). We used the training data to build the SVR models. Then, we predicted the test set using the resulting models.

5.3 Tuning Parameters

We propose to use RBF kernel function type in training tasks of kSVR, krSVR and LibSVM for building SVR models because it is general and efficient [24]. The cross-validation protocol (2-fold) is used to tune the regression tolerance ε, the hyper-parameter γ of RBF kernel (RBF kernel of two individuals x_i, x_j, $K[i,j] = exp(-\gamma\|x_i - x_j\|^2)$) and the cost C (a trade-off between the margin size and the errors) to obtain a

Table 1. Description of datasets

ID	Datasets	#Datapoints	#Dimensions	Target domain	Evaluation protocol
1	Appliances energy prediction	19 735	27	[10.0, 1080.0]	13 500 Trn - 6 235 Tst
2	Facebook comment volume	40 949	53	[0.0, 1 305.0]	27 500 Trn - 13 449 Tst
3	BlogFeedback	60 021	280	[0.0, 1424.0]	52 397 Trn - 7 624 Tst
4	Buzz in social media (Twitter)	583 250	77	[0.0, 75 724.5]	400 000 Trn - 183 250 Tst
5	YearPredictionMSD	515 345	90	[1 922, 2 011]	400 000 Trn - 115 345 Tst

good correctness. For two largest datasets (Buzz in social media Twitter, YearPrediction-MSD), we used a subset randomly sampling about 5% training dataset for tuning hyper-parameters due to the expensive computational cost.

Our kSVR uses the parameter k local models (number of clusters). We propose to set k so that each cluster consists of about 200 individuals. The idea gives a trade-off between the generalization capacity [15] and the computational cost. Furthermore, the krSVR algorithm learns 20 random k local SVR models ($T = 20$) with the number of random dimensions being the square root of the full set ($n' = \sqrt{(n)}$ as recommended by [20]).

Table 2 presents the hyper-parameters of kSVR, krSVR and LibSVM in the regression.

Table 2. Hyper-parameters of kSVR, krSVR and LibSVM

ID	Datasets	γ	C	ε	k
1	Appliances energy prediction	0.02	100 000	0.1	30
2	Facebook comment volume	0.001	100 000	0.1	300
3	BlogFeedback	0.4	100 000	0.05	500
4	Buzz in social media (Twitter)	0.1	100 000	0.1	4000
5	YearPredictionMSD	0.01	100 000	0.1	1500

5.4 Regression Results

The regression results of LibSVM, kSVR and krSVR on the five datasets are given in Table 3, Figs. 6, 7 and 8.

Table 3. Regression results in terms of mean absolute error (MAE) and training time (minutes)

		Mean absolute error (MAE)			Training time (min)		
ID	Datasets	LibSVM	kSVR	krSVR	LibSVM	kSVR	krSVR
1	Appliances energy prediction	47.81	47.94	45.92	2.55	0.05	0.22
2	Facebook comment volume	8.97	8.59	7.38	27.91	0.1	0.27
3	BlogFeedback	9.85	6.40	6.20	53.78	3.86	20.75
4	Buzz in social media (Twitter)	235.25	46.73	45.27	5193.59	31.94	133.07
5	YearPredictionMSD	8.18	7.86	7.82	2477.91	6.33	24.67
6	Average	62.01	23.50	22.52	1551.15	8.45	35.79

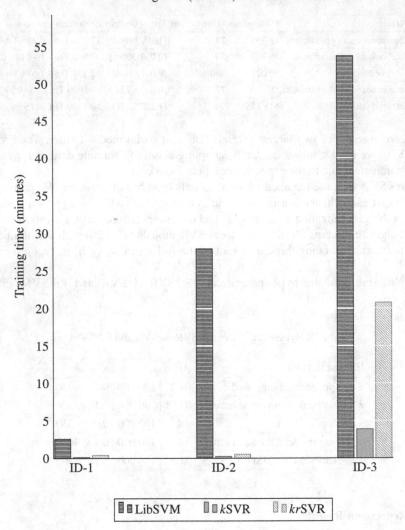

Fig. 6. Comparison of training time (minutes) for small datasets

Comparison in Training Time

As it was expected, our proposed algorithms of local SVR models (kSVR, krSVR) outperform LibSVM in terms of training time. The average of kSVR, krSVR and LibSVM training time are 8.45 min, 35.79 min and 1551.15 min, respectively. It means that kSVR is the fastest among the three learning algorithms. kSVR is 183.5 times faster than LibSVM. krSVR is also 43.34 times faster than LibSVM. Training time of krSVR is about 4.23 times longer than kSVR.

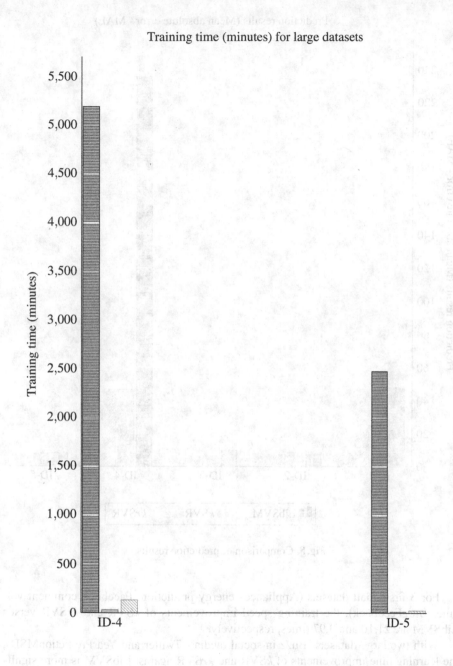

Fig. 7. Comparison of training time (minutes) for large datasets

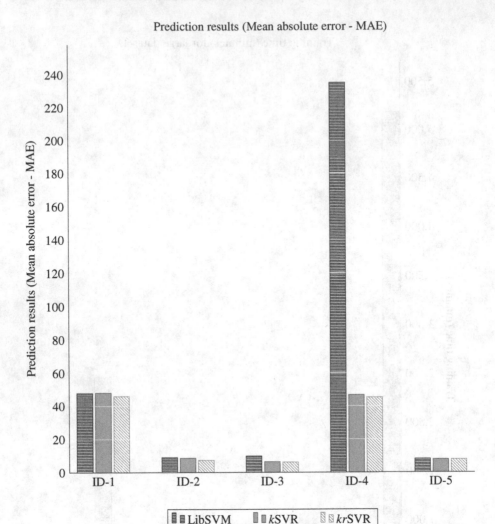

Fig. 8. Comparison of prediction results

For 3 first small datasets (Appliances energy prediction, Facebook comment volume, BlogFeedback), the training speed improvements of kSVR and krSVR versus LibSVM are 21.10 and 3.97 times, respectively.

With two large datasets (Buzz in social media - Twitter and YearPredictionMSD), the learning time improvements of kSVR and krSVR against LibSVM is more significant. These training speed improvements of kSVR and krSVR compared to LibSVM are 200.44 times and 48.63 times, respectively.

Comparison in Prediction Correctness
In terms of prediction correctness (measured by the mean absolute error - MAE), the error average made by kSVR, krSVR and LibSVM are 23.50, 22.52 and 62.01, respectively. The comparison of prediction correctness, dataset by dataset, shows that:

- kSVR has 4 wins, 1 defeat against LibSVM; kSVR is beaten only once (with Appliances energy prediction dataset) by LibSVM;
- krSVR is most accurate with 5/5 wins versus LibSVM and kSVR.

The regression results show that our kSVR, krSVR are more accurate than LibSVM for the prediction. It seems to be that local SVR models are suited for large datasets. The data partition stage allows local SVR algorithms being easily to fit regression models for data subsets. It leads to improve the prediction correctness of local SVR algorithms. In particular with a large dataset as Buzz in social media - Twitter, the target domain is very large $[0, 75724.5]$, therefore the prediction error made by any algorithm is also high. Our kSVR, krSVR can reduce the prediction error about 5 times compared to LibSVM.

The numerical results demonstrate that kSVR, krSVR improve not only the training time, but also the prediction correctness when dealing with large datasets. The regression results allow to believe that our proposed kSVR, krSVR are efficient for handling such large datasets.

6 Discussion on Related Works

Our proposal is related to large-scale SVM learning algorithms. The improvements of SVM training on very large datasets include effective heuristic methods in the decomposition of the original quadratic programming into series of small problems [10, 25, 26] and [6]. Recent works [27, 28] proposed the stochastic gradient descent methods for dealing with large scale linear SVM solvers. The parallel and distributed algorithms [29–31] for the linear classification improve learning performance for large datasets by dividing the problem into sub-problems that execute on large numbers of networked PCs, grid computing, multi-core computers.

The review paper [32] provides a comprehensive survey on large-scale linear support vector classification. LIBLINEAR [33] and its extension [34] uses the Newton method for the primal-form of SVM and and the coordinate descent approach for the dual-form SVM to deal with very large linear classification and regression. The parallel algorithms of logistic regression and linear SVM using Spark [35] are proposed in [36]. The distributed Newton algorithm of logistic regression [37] is implemented in the Message Passing Interface (MPI). The parallel dual coordinate descent method for linear classification [38] is implemented in multi-core environments using OpenMP. The incremental and decremental algorithms [39] aim at training linear classification of large data that cannot fit in memory. These algorithms are proposed to efficiently deal large-scale linear classification tasks in a very-high-dimensional input space. But the computational cost of a non-linear SVM approach is still impractical. The work in [40] tries to automatically determine which kernel classifiers perform strictly better than linear for a given data set.

Our proposal is in some aspects related to local SVM learning algorithms. The first approach is to classify data in hierarchical strategy. This kind of training algorithm performs the classification task with two main steps. The first one is to cluster the full dataset into homogeneous groups (clusters) and then the second one is to learn the local supervised classification models from clusters. The paper [41] proposed to use the expectation-maximization (EM) clustering algorithm [42] for partitioning the training set into k joint clusters (the EM clustering algorithm makes a soft assignment based on the posterior probabilities [43]); for each cluster, a neural network (NN) is learnt to classify the individuals in the cluster. The parallel mixture of SVMs algorithm proposed by [44] constructs local SVM models instead of NN ones in [41]. CSVM [45] uses $k$means algorithm [7] to partition the full dataset into k disjoint clusters; then the algorithm learns weighted local linear SVMs from clusters. More recent kSVM [46], krSVM [47] (random ensemble of kSVM), tSVM [16] propose to parallely train the local non-linear SVMs instead of weighting linear ones of CSVM. DTSVM [48,49] uses the decision tree algorithm [50,51] to split the full dataset into disjoint regions (tree leaves) and then the algorithm builds the local SVMs for classifying the individuals in tree leaves. These algorithms aim at speeding up the learning time.

The second approach is to learn local supervised models from k nearest neighbors (kNN) of a new testing individual. First local learning algorithm of Bottou & Vapnik [14] find kNN of a test individual; train a neural network with only these k neighborhoods and apply the resulting network to the test individual. k-local hyperplane and convex distance nearest neighbor algorithms are also proposed in [52]. More recent local SVM algorithms aim to use the different methods for kNN retrieval; including SVM-kNN [53] trying different metrics, ALH [54] using the weighted distance and features, FaLK-SVM [55] speeding up the kNN retrieval with the cover tree [56].

A theorical analysis for such local algorithms discussed in [13] introduces the tradeoff between the capacity of learning system and the number of available individuals. The size of the neighborhoods is used as an additional free parameters to control generalisation capacity against locality of local learning algorithms.

7   Conclusion and Future Works

We presented new parallel algorithms of local SVR that achieve high performances for the non-linear regression on large datasets. The training task of kSVR is to split the full training dataset into k clusters. This step is to reduce data size in training local SVR. And then it easily learns k non-linear SVR models in the parallel way on multi-core computers in which a SVR model is trained in each cluster to predict the data locally. The krSVR learning algorithm is to improve the generalization capacity of the kSVR alone by training an ensemble random kSVR models. The performance analysis and the numerical test results on five datasets from UCI repository showed that our proposed kSVR, krSVR are efficient in terms of training time and prediction correctness compared to the standard SVR. The learning time improvements of kSVR and krSVR versus LibSVM are 183.5 and 43.3 times. kSVR and krSVR improve 62.10%, 63.70% of the relative prediction correctness compared to the standard SVR, respectively. An example of kSVR's effectiveness is given with the non-linear regression of YearPredictionMSD dataset (having 400000 datapoints, 90 dimensions) in 6.33 min and 7.86 mean absolute error obtained on the prediction of the testset.

In the near future, we intend to provide more empirical test on large benchmarks and comparisons with other algorithms. A promising avenue for future research is able to automatically tune hyperparameters for kSVR and krSVR.

References

1. Lyman, P., et al.: How much information (2003)
2. National Research Council, Division on Engineering and Physical Sciences, Board on Mathematical Sciences and Their Applications, Committee on the Analysis of Massive Data, Committee on Applied and Theoretical Statistics: Frontiers in Massive Data Analysis. The National Academies Press (2013)
3. Vapnik, V.: The Nature of Statistical Learning Theory. Springer, Heidelberg (1995). https://doi.org/10.1007/978-1-4757-3264-1
4. Guyon, I.: Web page on SVM applications (1999). http://www.clopinet.com/isabelle/Projects/-SVM/app-list.html
5. Bui, L.D., Tran-Nguyen, M.T., Kim, Y.G., Do, T.N.: Parallel algorithm of local support vector regression for large datasets. In: Proceedings of Future Data and Security Engineering - 4th International Conference, FDSE 2017, pp. 139–153, Ho Chi Minh City, Vietnam, 29 November–1 December (2017)
6. Chang, C.C., Lin, C.J.: LIBSVM : a library for support vector machines. ACM Trans. Intell. Syst. Technol. **2**(27), 1–27 (2011)
7. MacQueen, J.: Some methods for classification and analysis of multivariate observations. In: Proceedings of 5th Berkeley Symposium on Mathematical Statistics and Probability, vol. 1, pp. 281–297. University of California Press, Berkeley, January 1967
8. Lichman, M.: UCI machine learning repository (2013)
9. Cristianini, N., Shawe-Taylor, J.: An Introduction to Support Vector Machines: And Other Kernel-based Learning Methods. Cambridge University Press, New York (2000)
10. Platt, J.: Fast training of support vector machines using sequential minimal optimization. In: Schölkopf, B., Burges, C., Smola, A. (eds.) Advances in Kernel Methods - Support Vector Learning, pp. 185–208 (1999)
11. OpenMP Architecture Review Board: OpenMP application program interface version 3.0 (2008)
12. Bi, J., Bennett, K.P.: A geometric approach to support vector regression. Neurocomputing **55**(1–2), 79–108 (2003)
13. Vapnik, V.: Principles of risk minimization for learning theory. In: Advances in Neural Information Processing Systems 4, NIPS Conference, Denver, Colorado, USA, 2–5 December 1991, pp. 831–838 (1991)
14. Bottou, L., Vapnik, V.: Local learning algorithms. Neural Comput. **4**(6), 888–900 (1992)
15. Vapnik, V., Bottou, L.: Local algorithms for pattern recognition and dependencies estimation. Neural Comput. **5**(6), 893–909 (1993)
16. Do, T.N., Poulet, F.: Parallel learning of local SVM algorithms for classifying large datasets. T. Large-Scale Data-Knowl.-Cent. Syst. **31**, 67–93 (2016)
17. Do, T.N., Poulet, F.: Latent-lSVM classification of very high-dimensional and large-scale multi-class datasets. Concurr. Comput.: Pract. Exp. **0**(0), e4224

18. Vapnik, V.: The Nature of Statistical Learning Theory, 2nd edn. Springer, Heidelberg (2000). https://doi.org/10.1007/978-1-4757-3264-1
19. Breiman, L.: Bagging predictors. Mach. Learn. **24**(2), 123–140 (1996)
20. Breiman, L.: Random forests. Mach. Learn. **45**(1), 5–32 (2001)
21. Breiman, L.: Arcing classifiers. Ann. Stat. **26**(3), 801–849 (1998)
22. Dietterich, T.G.: Ensemble methods in machine learning. In: Kittler, J., Roli, F. (eds.) MCS 2000. LNCS, vol. 1857, pp. 1–15. Springer, Heidelberg (2000). https://doi.org/10.1007/3-540-45014-9_1
23. Whaley, R., Dongarra, J.: Automatically tuned linear algebra software. In: Ninth SIAM Conference on Parallel Processing for Scientific Computing, CD-ROM Proceedings (1999)
24. Lin, C.: A practical guide to support vector classification (2003)
25. Boser, B., Guyon, I., Vapnik, V.: An training algorithm for optimal margin classifiers. In: Proceedings of 5th ACM Annual Workshop on Computational Learning Theory of 5th ACM Annual Workshop on Computational Learning Theory, pp. 144–152. ACM (1992)
26. Osuna, E., Freund, R., Girosi, F.: An improved training algorithm for support vector machines. In: Gile, L., Morgan, N., Wilson, E. (eds.) Neural Networks for Signal Processing VII, Jose Principe, pp. 276–285 (1997)
27. Shalev-Shwartz, S., Singer, Y., Srebro, N.: Pegasos: primal estimated sub-gradient solver for SVM. In: Proceedings of the Twenty-Fourth International Conference Machine Learning, pp. 807–814 (2007). ACM
28. Bottou, L., Bousquet, O.: The tradeoffs of large scale learning. In: Platt, J., Koller, D., Singer, Y., Roweis, S. (eds.) Advances in Neural Information Processing Systems, vol. 20, pp. 161–168. NIPS Foundation (2008). http://books.nips.cc
29. Do, T.N.: Parallel multiclass stochastic gradient descent algorithms for classifying million images with very-high-dimensional signatures into thousands classes. Vietnam. J. Comput. Sci. **1**(2), 107–115 (2014)
30. Do, T.N., Poulet, F.: Parallel multiclass logistic regression for classifying large scale image datasets. In: Advanced Computational Methods for Knowledge Engineering - Proceedings of 3rd International Conference on Computer Science, Applied Mathematics and Applications - ICCSAMA 2015, Metz, France, 11–13 May 2015, pp. 255–266 (2015)
31. Do, T.-N., Tran-Nguyen, M.-T.: Incremental parallel support vector machines for classifying large-scale multi-class image datasets. In: Dang, T.K., Wagner, R., Küng, J., Thoai, N., Takizawa, M., Neuhold, E. (eds.) FDSE 2016. LNCS, vol. 10018, pp. 20–39. Springer, Cham (2016). https://doi.org/10.1007/978-3-319-48057-2_2
32. Yuan, G., Ho, C., Lin, C.: Recent advances of large-scale linear classification. Proc. IEEE **100**(9), 2584–2603 (2012)
33. Fan, R.E., Chang, K.W., Hsieh, C.J., Wang, X.R., Lin, C.J.: LIBLINEAR: a library for large linear classification. J. Mach. Learn. Res. **9**(4), 1871–1874 (2008)
34. Ho, C., Lin, C.: Large-scale linear support vector regression. J. Mach. Learn. Res. **13**, 3323–3348 (2012)
35. Zaharia, M., Chowdhury, M., Franklin, M.J., Shenker, S., Stoica, I.: Spark: cluster computing with working sets. In: Proceedings of the 2nd USENIX Conference on Hot Topics in Cloud Computing, HotCloud 2010, p. 10. USENIX Association, Berkeley (2010)
36. Lin, C., Tsai, C., Lee, C., Lin, C.: Large-scale logistic regression and linear support vector machines using spark. In: 2014 IEEE International Conference on Big Data, Big Data 2014, Washington, DC, USA, 27–30 October 2014, pp. 519–528 (2014)
37. Zhuang, Y., Chin, W., Juan, Y., Lin, C.: Distributed Newton methods for regularized logistic regression. In: Proceedings Advances in Knowledge Discovery and Data Mining - 19th Pacific-Asia Conference, PAKDD 2015, Part II, Ho Chi Minh City, Vietnam, 19–22 May 2015, pp. 690–703 (2015)

38. Chiang, W., Lee, M., Lin, C.: Parallel dual coordinate descent method for large-scale linear classification in multi-core environments. In: Proceedings of the 22nd ACM SIGKDD International Conference on Knowledge Discovery and Data Mining, San Francisco, CA, USA, 13–17 August 2016, pp. 1485–1494 (2016)

39. Tsai, C., Lin, C., Lin, C.: Incremental and decremental training for linear classification. In: The 20th ACM SIGKDD International Conference on Knowledge Discovery and Data Mining, KDD 2014, New York, NY, USA, 24–27 August 2014, pp. 343–352 (2014)

40. Huang, H., Lin, C.: Linear and kernel classification: when to use which? In: Proceedings of the SIAM International Conference on Data Mining 2016 (2016)

41. Jacobs, R.A., Jordan, M.I., Nowlan, S.J., Hinton, G.E.: Adaptive mixtures of local experts. Neural Comput. **3**(1), 79–87 (1991)

42. Dempster, A.P., Laird, N.M., Rubin, D.B.: Maximum likelihood from incomplete data via the EM algorithm. J. R. Stat. Soc. Ser. B **39**(1), 1–38 (1977)

43. Bishop, C.M.: Pattern Recognition and Machine Learning. Springer, New York (2006)

44. Collobert, R., Bengio, S., Bengio, Y.: A parallel mixture of SVMs for very large scale problems. Neural Comput. **14**(5), 1105–1114 (2002)

45. Gu, Q., Han, J.: Clustered support vector machines. In: Proceedings of the Sixteenth International Conference on Artificial Intelligence and Statistics, AISTATS 2013, Scottsdale, AZ, USA, 29 April–1 May 2013, Volume 31 of JMLR Proceedings, pp. 307–315 (2013)

46. Do, T.-N.: Non-linear classification of massive datasets with a parallel algorithm of local support vector machines. In: Le Thi, H.A., Nguyen, N.T., Do, T.V. (eds.) Advanced Computational Methods for Knowledge Engineering. AISC, vol. 358, pp. 231–241. Springer, Cham (2015). https://doi.org/10.1007/978-3-319-17996-4_21

47. Do, T.-N., Poulet, F.: Random local SVMs for classifying large datasets. In: Dang, T.K., Wagner, R., Küng, J., Thoai, N., Takizawa, M., Neuhold, E. (eds.) FDSE 2015. LNCS, vol. 9446, pp. 3–15. Springer, Cham (2015). https://doi.org/10.1007/978-3-319-26135-5_1

48. Chang, F., Guo, C.Y., Lin, X.R., Lu, C.J.: Tree decomposition for large-scale SVM problems. J. Mach. Learn. Res. **11**, 2935–2972 (2010)

49. Chang, F., Liu, C.C.: Decision tree as an accelerator for support vector machines. In: Ding, X. (ed.) Advances in Character Recognition. InTech (2012)

50. Quinlan, J.R.: C4.5: Programs for Machine Learning. Morgan Kaufmann, San Mateo (1993)

51. Breiman, L., Friedman, J.H., Olshen, R.A., Stone, C.: Classification and Regression Trees. Wadsworth International, Kennett Square (1984)

52. Vincent, P., Bengio, Y.: K-local hyperplane and convex distance nearest neighbor algorithms. In: Advances in Neural Information Processing Systems, pp. 985–992. The MIT Press (2001)

53. Zhang, H., Berg, A., Maire, M., Malik, J.: SVM-KNN: discriminative nearest neighbor classification for visual category recognition. In: 2006 IEEE Computer Society Conference on Computer Vision and Pattern Recognition, vol. 2, pp. 2126–2136 (2006)

54. Yang, T., Kecman, V.: Adaptive local hyperplane classification. Neurocomputing **71**(13–15), 3001–3004 (2008)

55. Segata, N., Blanzieri, E.: Fast and scalable local kernel machines. J. Mach. Learn. Res. **11**, 1883–1926 (2010)

56. Beygelzimer, A., Kakade, S., Langford, J.: Cover trees for nearest neighbor. In: Proceedings of the 23rd International Conference on Machine Learning, pp. 97–104. ACM (2006)

A Parallel Incremental Frequent Itemsets Mining IFIN⁺: Improvement and Extensive Evaluation

Van Quoc Phuong Huynh[1(✉)], Josef Küng[1], and Tran Khanh Dang[2]

[1] Institute for Application Oriented Knowledge Processing (FAW),
Faculty of Engineering and Natural Sciences (TNF),
Johannes Kepler University (JKU), Linz, Austria
{vqphuynh, jkueng}@faw.jku.at
[2] Ho Chi Minh City University of Technology, VNUHCM,
Ho Chi Minh City, Vietnam
khanh@hcmut.edu.vn

Abstract. In this paper, we propose a shared-memory parallelization solution for the Frequent Itemsets Mining algorithm IFIN, called IFIN⁺. The motivation for our work is that commodity processors, nowadays, are enhanced with many physical computational units, and exploiting full advantage of this is a potential solution to improve computational performance in single-machine environments. The portions in the serial version are improved in means which increases efficiency and computational independence for convenience in designing parallel computation with Work-Pool model, be known as a good model for load balance. We conducted extensive experiments on both synthetic and real datasets to evaluate IFIN⁺ against its serial version IFIN, the well-known algorithm FP-Growth and other two state-of-the-art ones, FIN and PrePost⁺. The experimental results show that the running time of IFIN⁺ is the most efficient, especially in the case of mining at different support thresholds within the same running session. Compare to its serial version, IFIN⁺ performance is improved significantly.

Keywords: Incremental · Parallel · Frequent Itemsets Mining · Data mining · Big Data · IPPC-Tree · IFIN · IFIN⁺

1 Introduction

Frequent Itemsets Mining (FIM) can be briefly described as follows. Given a dataset of n transactions $D = \{T_1, T_2, \ldots, T_n\}$, the dataset contains a set of m distinct items $I = \{i_1, i_2, \ldots, i_m\}$, $T_i \subseteq I$. A k-itemset, IS, is a set of k items ($1 \leq k \leq m$). Each itemset IS possesses an attribute, *support*, which is the number of transactions containing IS. FIM is featured by a support threshold ε which is the percent of transactions in the whole dataset D. An itemset IS is called frequent itemset iff $IS.support \geq \varepsilon * n$. The problem is to discover all frequent itemsets existing in D.

Discovering frequent itemsets in a large dataset is an important problem in data mining. In Big Data era, this problem, as well as other mining models, has been being

© Springer-Verlag GmbH Germany, part of Springer Nature 2019
A. Hameurlain et al. (Eds): TLDKS XLI, LNCS 11390, pp. 78–106, 2019.
https://doi.org/10.1007/978-3-662-58808-6_4

challenged by very large volume and high velocity of datasets which are fast accumulated over time. Fortunately, nowadays, RAM memory has a larger capacity and becomes much cheaper, and commodity processors' computational power is enhanced considerably with many physical computational units. To take this advantage and confront with the challenge, we propose an algorithm, named IFIN⁺, as a solution for parallelizing our previous work IFIN [18] (Incremental Frequent Itemsets Nodesets) algorithm with shared-memory multithreads. The purpose is to improve the performance of IFIN by enhancing the computational efficiency and independence and increasing the throughput in single-machine environments. In general, IFIN algorithm encompasses four phases: (1) IPPC-Tree (Incremental Pre-Post-Order Coding Tree) construction, (2) Frequent 2-itemsets generation, (3) Nodesets for frequent 2-itemsets generation, (4) Frequent k-itemsets generation $(k > 2)$. These four phases can be divided into small independent chunks of work and processed separately by more than one thread. The synchronization at the end of a phase will delay the start of the next one and result in longer mining time if load balance is not guaranteed. To avoid this problem, therefore, all these four processing phases are designed in Work-Pool model, a well-known model for load balance, in which all workers continuously fetch and process small chunks of work until there are no more tasks in the work pool. In the first phase, a new efficient stored format and parallel loading for the IPPC-Tree are proposed. The second and third phases are changed to increase the computational independence for parallelization; and the last one, frequent k-itemsets generation $(k > 2)$, is parallelized as well. By that solution, the running time of IFIN⁺ is improved significantly compared to its serial version IFIN. This paper is the next work upon our previous improvement [19] for IFIN. In that, the new efficient stored format for IPPC-Tree and parallelization for the four processing phase are introduced; and extensive experiments are conducted on both synthetic and real datasets to evaluate the performance of IFIN⁺ against its serial version IFIN, the well-known algorithm FP-Growth and other two state-of-the-art ones FIN and PrePost⁺.

The rest of the paper is organized as follows. In Sect. 2, some related works are presented. Section 3 introduces the IPPC-Tree structure and stored formats, some relevant algorithms and solutions for loading the IPPC-Tree. The algorithm IFIN⁺ is mentioned in Sect. 5 based on preliminaries in Sect. 4 and followed with experiments in Sect. 6. Finally, conclusions are given in Sect. 7.

2 Related Works

The problem of mining frequent itemsets was started up by Agrawal & Srikant with algorithm Apriori [1]. This algorithm generates candidate $(k + 1)$-itemsets from frequent k-itemsets at the $(k + 1)^{\text{th}}$ pass and then scans dataset to check whether a candidate $(k + 1)$-itemsets is a frequent one. Many previous works were inspired by this algorithm. Algorithm Partition [8] aims at reducing I/O cost by dividing a dataset into non-overlapping and memory-fitting partitions which are sequentially scanned in two phases. In the first phase, local candidate itemsets are generated for each partition, and then they are checked in the second one. DCP [9] enhances Apriori by incorporating two dataset pruning techniques introduced in DHP [10] and using direct counting method for storing candidate itemsets and counting their support. In general, Apriori-like methods suffer from two drawbacks: a deluge of generated candidate itemsets and/or I/O overhead caused by repeatedly scanning dataset. Two other approaches, which are more efficient than Apriori-like methods, are also proposed to solve the problem: (1) frequent pattern growth adopting divide-and-conquer with FP-Tree structure and FP-Growth [2], and (2) vertical data format strategy in Eclat [11]. FP-Growth and algorithms based on it such as [12, 13] are efficient solutions since unlike Apriori, they avoid many times of scanning dataset and generation-and-test. However, they become less efficient when datasets are sparse. While algorithms based on FP-Growth and Apriori use a horizontal data format; Eclat and some other algorithms [8, 14, 15] apply vertical data format, in which each item is associated with a set of transaction identifiers, Tids, containing the item. This approach avoids scanning dataset repeatedly, but a huge memory overhead is expensed for sets of Tids when dataset becomes large and/or dense. Recently, two remarkably efficient algorithms are introduced: FIN [4] with POC-Tree and PrePost$^+$ [5] with PPC-Tree. These two structures are prefix trees and similar to FP-Tree, but the two mining algorithms use additional data structures, called Nodeset and N-list respectively, to significantly improve mining speed.

To better deal with the challenge of high volume in Big Data, in addition to the ideas of parallel mining for existing algorithms such as [16] for Eclat, incremental mining approaches are also considered as a potential solution. Some typical algorithms in this approach are algorithm FELINE [3] with CATS-Tree structure and IM_WMFI [17] for mining weighted maximal frequent itemsets from incremental datasets. These methods are both based on the well-known FP-Tree for its efficiency.

3   IPPC Tree

IPPC-Tree is a prefix tree and possesses two properties, Properties 1 and 2. IPPC-Tree includes one root node labeled "*root*" and a set of prefix subtrees as its children. Each node in the subtrees contains the following attributes:

- *item-name*: the name of an item in a transaction that the node registered.
- *support* (or *local support* of an item): the number of transactions containing the node's *item-name*. Conversely, *global support* of an item, without concerning nodes, is the number of transactions containing the item.
- *pre-order* and *post-order*: two global identities in the IPPC-Tree which are sequent numbers generated by traversing the tree with pre and post order respectively.

Property 1: For a given IPPC-Tree, there exist no duplication nodes with the same item in a path of nodes from the root to a leaf node.

Property 2: In a given IPPC-Tree, the *support* of a parent node must be greater than or equal to the sum of all its children's *support*.

IPPC-Tree is a combination of (1) the idea of the flexible and local order of items in a path from the root to a leaf node in CATS-Tree [3] and (2) the PPC-Tree [5] which each node in PPC-Tree is identified by a pair of codes: *pre-order* and *post*-order. The construction of the IPPC-Tree does not require a given support threshold. The tree is a compact and information-lossless structure of the whole items of all transactions in a given dataset. Local order of items in a path of nodes from the root to a leaf is flexible and can be changed to improve compression while remaining Property 2. To guarantee this, two conditions for swapping are as follows.

Child Swapping: A node can be swapped with its child node if it has only one child node, its *support* is equal to its child's *support*, and the number of child nodes of its child is not greater than one.

Descendant Swapping: Given a path of k nodes $N_1 \rightarrow N_2 \rightarrow \cdots \rightarrow N_k (k > 2)$, N_i is the parent node of $N_j (i < j)$; if every node $N_i (i < k)$ satisfies the Child Swapping condition, node N_1 can be swapped with descendant node N_k.

For an overview, Table 1 provides a comparison among the four similarity tree structures FP-Tree, CATS-Tree, PPC-Tree, and IPPC-Tree.

Table 1. Comparison among FP-Tree, CATS-Tree, PPC-Tree, and IPPC-Tree

	FP-Tree	CATS-Tree	PPC-Tree	IPPC-Tree
Items building the tree	Frequent items	All items	Frequent items	All items
Node attributes	- Item-name - Support	- Item-name - Support	- Item-name - Support - Pre-order - Post-order	- Item-name - Support - Pre-order - Post-order
Header table and node chains of the same item	Yes	Yes	No	No
Local order of items in a path (from the root to a leave) is based on	Global support	Local support	Global support	Local support
Local order of items in a path is flexible	No	Yes	No	Yes
Order of child nodes with the same parent node	No	Descending of support	No	No

To demonstrate the building process of an IPPC-Tree, the Fig. 1 records transaction by transaction in Table 1 inserted into an empty IPPC-Tree. Initially, the tree has only the root node, and transaction $1(b, e, d, f, c)$ is inserted as it is in Fig. 1(a). The Fig. 1(b) is of the tree after transaction $2(d, c, b, g, f, h)$ is added. The item b in transaction 2 is merged with node b in the tree. Although transaction 2 does not contain item e, but its common items $d, f,$ and c can be merged with the corresponding nodes. Item d is found common, so it is merged with node d after node d is swapped[1] with node e to guarantee the Property 2. Similarly, items f and c are merged with node f and c respectively; and the remaining items g and h are inserted as a child branch of node c. In Fig. 1(c),

---

[1] Swapping two nodes is simply exchanging one's item name to that of the other.

transaction 3(*f*, *a*, *c*) is processed. Common item *f* is found that can be merged with node *f*, so node *f* is swapped with node *b*. Item *c* is also a common one, but it is not able to be merged with node *c* as node *d* does not satisfy the Descendant Swapping condition with node *c*. Then the items *a* and *c* are added as a branch from node *f*. When transaction 4(*a*, *b*, *d*, *f*, *c*, *h*) is added in Fig. 1(d), common items *f*, *d*, *b*, and *c* are merged straightforwardly with corresponding nodes *f*, *d*, *b*, and *c*. The remaining items *a* and *h* are then inserted into the subtree having root node *c*. The item *h* is found common with node *h* in the second branch. Node *h* and item *h*, therefore, are merged together after node *h* is swapped with node *g*. The last item *a* is then inserted as a new child branch from node *h*. Insertion of transaction 5(*b*, *d*, *c*) is depicted in Fig. 1(e). All items in transaction 5 are common, but they cannot be merged with nodes *b*, *d*, and *c* as node *f* does not guarantee the Child Swapping condition. Thus, transaction 5 is added as a new child branch of the root node.

After the dataset has been processed, each node in the IPPC-Tree is attached with a pair of sequent numbers (*pre-order*, *post-order*) by scanning the tree with pre-order and post-order traversals through procedure **AssignPrePostOrder**. For an example, node (4, 6) is identified by *pre-order* = 4 and *post-order* = 6, and it registers item *b* with *support* = 3. The above are all concepts of the IPPC-Tree construction; for a formal and detail description, the Algorithm 1 for building the tree, **BuildIPPCTree**, is presented as follows.

Table 2. Example transaction dataset

ID	Items in transactions
1	b, e, d, f, c
2	d, c, b, g, f, h
3	f, a, c
4	a, b, d, f, c, h
5	b, d, c

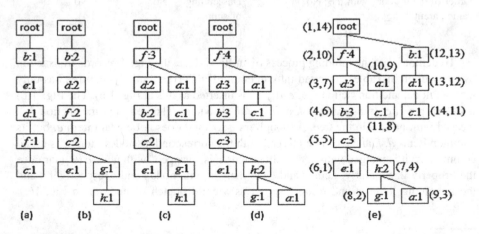

Fig. 1. An illustration for constructing an IPPC-Tree on example transaction dataset

Algorithm 1: BuildIPPCTree
Input: Dataset *D*, root node *R*
Output: An IPPC-Tree with root *R*, item list *L*
1. **For Each** transaction *T* ∈ *D*
2.   Update items and their supports in *L* from items in *T*;
3.   **InsertTransaction**(*T*, *R*);
4. **End For**
5. **AssignPrePostOrder**(*R*);

Procedure InsertTransaction(Transaction *T*, Node *R*)
1. *subNode* ← *R*; *notMerged*;
2. **While**(*T* ≠ ∅)
3.   *notMerged* ← **true**;
4.   **For Each** child node *N* of *subNode*
5.   **If**(*N.item-name* ∈ *T*)
6.   *notMerged* ← **false**; *N.support*++;
7.   *subNode* ← *N*; *T* ← (*T* \ *N.item-name*); **break**;
8.   **End If**
9.   **End For**
10.   **If**(*notMerged*) **break**;
11. **End While**
12. **If**(*T* = ∅) **Return**;
13. **For Each** child node *N* of *subNode*
14.   **If**(**MergeDescendants**(*T*, *N*)) **Return**;
15. **End For**
16. Insert *T* as a new branch from *subNode* (added nodes are initialized at 1 for their supports);

Function MergeDescendants(Transaction *T*, Node *N*)
1. *subNode* ← *N*; *mrgNode* ← *N*; *merged* ← **false**;
2. **While**(*subNode* satisfies the **Child Swapping** condition)
3.   *descendant* ← *subNode.child*;
4.   **If**(*descendant.item-name* ∈ *T*)
5.   *T* ← (*T* \ *descendant.item-name*); *merged* ← **true**;
6.   Exchange item names of *mrgNode* and *descendant*;
7.   *mrgNode.support*++; *mrgNode* ← *mrgNode.child*;
8.   **End If**
9.   *subNode* ← *descendant*;
10. **End While**
11. **If**(*merged*) Insert *T* as a new branch from *mrgNode.parent* (added nodes are initialized at 1 for their supports);
12. **Return** *merged*;

Procedure AssignPrePostOrder (Node *R*)
  // *PreOrder* and *PostOrder* are initialized at 1.
1. *R.pre-order* ← *PreOrder*; *PreOrder*++;
2. **For Each** child node *N* of *R* **Do AssignPrePostCode**(*N*);
3. *R.post-order* ← *PostOrder*; *PostOrder*++;

The IPPC-Tree construction requires only one dataset scanning and is independent of the support threshold as well as the global order of items (based on items' global supports) in a dataset. The tree is a compact and information-lossless structure of the whole items from all transactions in a given dataset. Therefore, a built IPPC-Tree from a dataset D is mined at different support thresholds and reused to build up a new IPPC-Tree corresponding to a new dataset $D' = D + \Delta D$.

To complete providing the incremental ability for the IPPC-Tree, methods of storing and loading for the tree and item list $\mathcal{L}$ must be proposed, in which the data format and algorithms are their two features. For the simplicity of storing and loading for $\mathcal{L}$, this detail will not be mentioned here. Beside *item-name* and *support*, the important information for loading a node is its parent's information to identify where the node was in the built tree. Our previous works [18, 19] utilize the *pre-order* (or *post-order*), the global identity, to determine the parent node for a node. The expression (1) shows the data format for a single node. For more efficient in storing and loading the IPPC-Tree, we propose a new format for a single node at (2).

$$< parent's\ pre\text{-}order > \ : \ < pre\text{-}order > \ : \ < post\text{-}order > \ : \ < item\text{-}name > \ : \ < support >$$

(1)

$$< 1|0 > \ < i|o > \ < item\text{-}name > \ : \ < support >$$

(2)

Table 3. Stored data formats for IPPC-Tree

Stored format for IPPC-Tree	New stored format for IPPC-Tree
<No. of Transactions>	<No. of Transactions>
-1:1:14:root:0	<No. of Nodes>
1:2:10:f:4	0if:4
1:12:13:b:1	0ib:1
2:3:7:d:3	1id:3
2:10:9:a:1	1ia:1
12:13:12:d:1	0id:1
3:4:6:b:3	1ib:3
10:11:8:c:1	0oc:1
13:14:11:c:1	1oc:1
4:5:5:c:3	0ic:3
5:6:1:e:1	1oe:1
5:7:4:h:2	1ih:2
7:8:2:g:1	0og:1
7:9:3:a:1	0oa:1

We employ Breadth-First-Search traversal to store the IPPC-Tree. In fact, the storing phrase can utilize other strategies such as pre-order traversal, but the sequence of node records generated by Breadth-First-Search traversal is more convenient for the loading phase. The reason is that the records of all child nodes with the same parent node are continuous together. Utilizing this property, the first element <1|0> in the new format (2) is only a character with two possible values '1' and '0', but it contains enough information to separate node groups (all nodes with the same parent). The second element <i|o> indicates the corresponding node is an inner node with the character 'i' or a leaf node with the character 'o'. An inner node means that it is the parent node of a certain node group. The parent node of the first group is the root node; the parent node of the second group is the first inner node; the parent node of the third one is the second inner node; and so on by that order all nodes will be placed at the right positions where they were in the built tree.

By storing the data record of every single node on a line, the corresponding data for the example tree in Fig. 1(e) with old format (1) and new format (2) is presented in the left and the right columns of Table 3 respectively. Obviously, the new format is more efficient than the old one in both aspects computation and storing volume. The algorithm for loading the IPPC-Tree in IFIN [18], using old stored format, is shown in procedure **LoadIPPCTree**.

```
Procedure LoadIPPCTree(File F, Root R, L, transCount)
1.   L ← Load the list of items;
2.   transCount ← Load the count of transactions;
3.   R ← Load the root node;
4.   parentNode ← R; nodeList ← ∅;
5.   For Each line L in data file F
6.      parentID ← extract <parent's pre-order> from L;
7.      Create a node N from L;
8.      Add N into the end of nodeList;
9.      While(parentID <> parentNode.pre-order){
10.        parentNode ← nodeList[0];
11.        Remove parentNode from nodeList;
12.     }
13.     Add N as a child of parentNode;
14.  End For
```

When the dataset becomes larger with the progress of additional data accumulated, the stored data for the built tree is also bigger; and the tree loading will consume more time as a result. Therefore, improving efficiency for procedure **LoadIPPCTree** is necessary. Loading the IPPC-Tree comprises three tasks for each line of data: (1) read a line, (2) parse the line and build a corresponding node, (3) connect the node to the tree. We realize that the second task takes most of the total time; and fortunately, the second task is performed in main memory. Therefore, it will be not interrupted by waiting for I/O when there is more than one thread shares the same I/O stream.

In this work, the parallelization for the IPPC-Tree loading is based on the new format (2). The parallel design is depicted in the Fig. 2. The file of a built IPPC-Tree is divided into n chunks of l lines and processed by k threads ($k \ll n$). The last chunk's number of lines may be lesser than l. Each time, a thread reads a chunk into its local buffer and sequentially creates a node for each data line. A shared reference array *FArray* is maintained for all created nodes, and a corresponding shared Boolean array

InnerNodePosition is used to indicate a node at a certain index in *FArray* is an inner node or not. The group node information for a node is temporarily recorded in *pre-order* property of the node. The connections among nodes in the IPPC-Tree will be established after node creation stage has finished. We can see that the access address spaces of individual threads in the *FArray* and *InnerNodePosition* are different. Hence, independence between threads is guaranteed. The parallelization is given in procedure **ParallelLoadIPPCTree**.

```
ParallelLoadIPPCTree(File F, Root R, L, transCount, threadCount)
1.   L ← Load list of items;
2.   transCount ← Load the count of transactions;
3.   nodeCount ← Load the count of nodes;
4.   Allocate FArray, InnerNodePositions with length nodeCount;
5.   index ← 0;
6.   For i From 1 To threadCount
7.     Start LoadingThread(F, FArray, InnerNodePosition, index);
8.   Synchronize all threads;
9.   parentIndex ← -1; parentNode ← R;
10.  groupId ← FArray[0].pre-order;
11.  For Each node N in FArray
12.    If(N.pre-order <> groupId){
13.      groupId ← N.pre-order;
14.      While(true) If(innerNodePosition[++parentIndex])break;
15.      parentNode ← FArray[parentIndex];
16.    }
17.    Add N as a child of parentNode;
18.  End For


LoadingThread(File F, FArray, InnerNodePosition, lineIndex)
1.   startIndex ← 0; lineCount ← 0;
2.   While(Work-Pool <> ∅)
3.     Mutually-exclusive-region {
4.       startIndex ← lineIndex;
5.       Load a chunk from F into Buffer;
6.       lineCount ← number of loaded lines;
7.       lineIndex ← lineIndex + lineCount;
8.     }
9.     For i From startIndex To (startIndex + lineCount - 1){
10.      Create a node N from the next line L in Buffer;
11.      FArray[i] ← N;
12.      N.pre-order ← (L[0] == '1' ? 1 : 0);
13.      InnerNodePosition[i] ← (L[1] == 'i');
14.    }
15.  End While
```

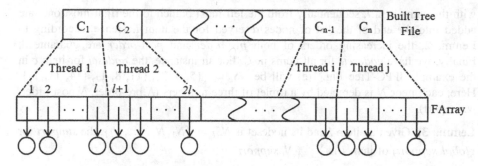

Fig. 2. The concept of parallelization for IPPC-Tree loading

The new stored format (2) does not contain *pre-order* and *post-order* information. Hence after loading completely the tree, procedure **AssignPrePostOrder** is executed. In case there is additional dataset, this procedure will be called after the tree is built up with the new dataset.

4 Preliminaries

In this subsection, some IPPC-Tree related definitions and lemmas are introduced as preliminaries for IFIN⁺ algorithm. For convenience in expressing the relative order between two items in an item list $\mathcal{L} = \{I_1, I_2, \ldots, I_n\}$, we denote $I_i \prec I_j$ to indicate that I_i is in front of I_j $(1 \leq i < j \leq n)$. There are two premises of traversing a tree with pre order and post order as follows:

Premise 1: Traversing a tree to process a work at each node with pre-order, it must be that (1) N_1 is an ancestor of N_2 or (2) N_1 and N_2 stay in two different branches (N_1 in the left and N_2 in the right) iff the work is done at N_1 before N_2.

Premise 2: Traversing a tree to process a work at each node with post order, it must be that (1) N_1 is an ancestor of N_2 or (2) N_1 and N_2 stay in two different branches (N_1 in the right and N_2 in the left) iff the work is done at N_2 before N_1.

By applying a work which assigns an increasingly global number at each node on Premise 1 & 2, two following lemmas are directly deduced.

Lemma 1: For any two different nodes N_1 and N_2 in the IPPC-Tree, N_1 is an ancestor of N_2 iff $N_1.pre\text{-}order < N_2.pre\text{-}order$ and $N_1.post\text{-}order > N_2.post\text{-}order$.

Lemma 2: For any two nodes N_1 and N_2 in two different branches of the IPPC-Tree, N_1 is in the left branch and N_2 in the right one iff $N_1.pre\text{-}order < N_2.pre\text{-}order$ and $N_1.post\text{-}order < N_2.post\text{-}order$.

Definition 1 (nodeset of an item): Given an IPPC-Tree, the *nodeset* of an item I, denoted by NS_I, is a set of all nodes in the IPPC-Tree with ascending order of *pre-order* and *post-order* in which all the nodes register the same item I.

In case N_1 and N_2 register the same item, N_1 and N_2 must be in two different branches because of Property 1. By traversing the IPPC-Tree with pre order, all nodes with the same item I, sequentially from the left-most branch to the right-most one, are added into the end of the list of nodes reserved for the item I. Hence, according to Lemma 2, the increasing orders of both *pre-order* and *post-order* are guaranteed. Finally, we have *nodesets* for all items in $\mathcal{L}$. For an instance, the *nodeset* for item c in the example IPPC-Tree Fig. 1(e) will be $NS_c = \{(5, 5, 3), (11, 8, 1), (14, 11, 1)\}$. Here, each node N is depicted by a triplet of three numbers (N.pre-order, N.post-order, N.support).

Lemma 3: Given an item I and its nodeset is $NS_I = \{N_1, N_2, \ldots, N_l\}$, the *support* (or *global support*) of item I is $\sum_{i=1}^{l} N_i.support$.

Rationale: According to Definition 1, NS_I includes all nodes registering the item I, and each node's *support* is the *local support* of the item I. Hence, the *global support* is the sum of all nodes' supports in NS_I. ∎

Definition 2 (nodeset of a k-itemset, $k \geq 2$): Given two $(k\text{-}1)$-itemsets $P_1 = p_1p_2\ldots p_{k-2}p_{k-1}$ with *nodesets* NS_{P_1} and $P_2 = p_1p_2\ldots p_{k-2}p_k$ with *nodeset* NS_{P_2} ($p_1 \prec p_2 \prec \ldots \prec p_k$), the *nodeset* of k-itemset $P = p_1p_2\ldots p_{k-2}p_{k-1}p_k$, NS_P, is defined as follows.

$$NS_P = \left\{ D_k \middle| \begin{array}{l} D_k = Descendant(N_i, M_j)\,with\,N_i \in NS_{P_1} \wedge M_j \in NS_{P_2} \\ D_k \in NS_{P_1} \wedge D_k \in NS_{P_2} \end{array} \right\}$$

Function $Descendant(N_i, M_j)$ means that there has been an ancestor-descendant relationship between N_i and M_j, and the output is the descendant node.

Lemma 4: Given a k-itemset P and its nodeset is $NS_P = \{N_1, N_2, \ldots, N_l\}$, the *support* of the itemset P is $\sum_{i=1}^{l} N_i.support$.

Proof. By the inductive method, the proof begins with *nodesets* for 2-itemsets.

According to the Definition 2, the *nodeset* of 2-itemset $p_1p_2(p_1 \prec p_2)$ is a set of all descendant nodes from all pairs of nodes $N_i \in NS_{p_1}$ and $M_j \in NS_{p_2}$ that both N_i and M_j stay in the same path of nodes from root to a leaf. Following the Property 2, descendant nodes' *supports* are lesser than or equal to that of the corresponding ancestor nodes; so the *supports* of descendant nodes are the local *supports* of all 2-itemsets p_1p_2 distributed in the IPPC-Tree. Consequently, Lemma 4 holds for case $k = 2$.

Assume Lemma 4 holds for case k. We need to proof Lemma 4 also holds for case $k + 1$. We have the assumptions:

1. k-itemset $P_1 = p_1p_2\ldots p_{k-1}p_k$ with its nodeset $NS_{P_1} = \{N_1, N_2, \ldots, N_{l1}\}$.
2. k-itemset $P_2 = p_1p_2\ldots p_{k-1}p_{k+1}$ with its nodeset $NS_{P_2} = \{M_1, M_2, \ldots, M_{l2}\}$.
3. For each N_i, there will be a path of k nodes from the root to N_i (except the root node); and N_i is the bottom node in the path visualized in Fig. 3(a). Each node in the path registers a certain item in $\{p_1, p_2, \ldots, p_{k-1}, p_k\}$. Each node M_j is similar to N_i, see Fig. 3(b).
4. $(k + 1)$-itemset $P = p_1p_2\ldots p_{k-1}p_kp_{k+1}$ with its nodeset $NS_P = \{D_1, D_2, \ldots, D_{l3}\}$.

According to the Definition 2, each D_i will fall into one of two cases:

Case 1 (Fig. 3(c)): $D_i = Descendant(N_i, M_j)$, without loss of generality, assume that N_i is the descendant node of M_j. In fact, two paths of nodes $1 \rightarrow 2 \rightarrow \cdots \rightarrow (k-1) \rightarrow k$ and $1' \rightarrow 2' \rightarrow \cdots \rightarrow (k-1)' \rightarrow k'$ must be in the same path from the root to a leave node as each child node has only one parent node. Because two paths of nodes share the common items in $\{p_1, p_2, \ldots, p_{k-1}\}$, and no duplicate nodes exist in the same path (Property 1); there will be $(k-1)$ pairs of identical nodes in which each pair includes one node from $1 \rightarrow 2 \rightarrow \cdots \rightarrow (k-1)$ and the other from $1' \rightarrow 2' \rightarrow \cdots \rightarrow (k-1)' \rightarrow k'$. This derives that there are only one path of $(k+1)$ unique nodes registering $(k+1)$ items in the item list $(p_1, p_2, \ldots, p_{k+1})$. Therefore, the *support* of N_i or D_i is a local *support* of the $(k+1)$-itemset $P = p_1 p_2 \ldots p_{k-1} p_k p_{k+1}$.

Case 2 (Fig. 3(d)): $N_i \equiv M_j$. This means that the items of N_i and M_j are the same and must register one in $\{p_1, p_2, \ldots, p_{k-1}\}$. Hence, $(k-2)$ remaining common items are shared by two paths of nodes $1 \rightarrow 2 \rightarrow \cdots \rightarrow (k-1)$ and $1' \rightarrow 2' \rightarrow \cdots \rightarrow (k-1)'$. By the same reasoning as in case 1, these two paths must be in the same path and there are $(k-2)$ pairs of identical nodes. Consequently, the number of unique nodes in the only node path is $(1 + (k-2) + 2 = k+1)$, and these $(k+1)$ nodes register $(k+1)$ items in the list of items $(p_1, p_2, \ldots, p_{k+1})$. Thus, the *support* of N_i or D_i is a local *support* of the $(k+1)$-itemset $P = p_1 p_2 \ldots p_{k-1} p_k p_{k+1}$.

Based on the two cases, Lemma 4 also holds for case $k+1$. Hence, Lemma 4 holds. ∎

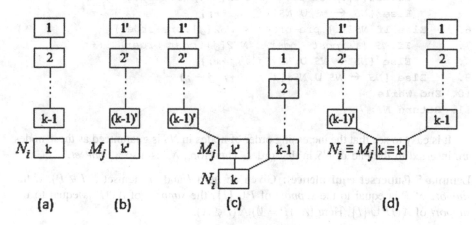

Fig. 3. Cases in definition 2

Given two $(k-1)$-itemsets $P_1 = p_1 p_2 \ldots p_{k-2} p_{k-1}$ and $P_2 = p_1 p_2 \ldots p_{k-2} p_k$ with their *nodesets* $NS_{P_1} = \{N_1, N_2, \ldots, N_{l1}\}$ and $NS_{P_2} = \{M_1, M_2, \ldots, M_{l2}\}$; at first glance, the computational complexity of generating *nodeset* NS_P for k-itemset $P = p_1 p_2 \ldots p_k$ is $O(l1 * l2)$. In fact, this complexity can be reduced significantly to

$O(l1 + l2)$, a linear cost, by utilizing Lemmas 1 and 2. For each pair of nodes N_i and M_j ($1 \le i \le l1, 1 \le j \le l2$), there are the following five cases:

1. (N_i.pre-order $> M_j$.pre-order) $\wedge$ (N_i.post-order $> M_j$.post-order): The relationship between N_i and M_j is not an ancestor-descendant relationship, so no node is added to NS_P. Certainly, M_j also does not have this relationship with remaining nodes in NS_{P_1} as increasing orders of both *pre-order* and *post-order* in *nodesets*. Therefore, M_{j+1} is selected as the next node for the next comparison.

2. (N_i.pre-order $> M_j$.pre-order) $\wedge$ (N_i.post-order $< M_j$.post-order): N_i is added to NS_P as N_i is the descendant node of M_j. Consequently, N_{i+1} is selected as the next node for the next comparison.

3. (N_i.pre-order $< M_j$.pre-order) $\wedge$ (N_i.post-order $> M_j$.post-order): Similar to the case 2, M_j is added to NS_P, and M_{j+1} is the next node for the next comparison.

4. (N_i.pre-order $< M_j$.pre-order) $\wedge$ (N_i.post-order $< M_j$.post-order): This case is similar to the case 1; and N_{i+1}, therefore, is the next node for the next comparison.

5. $N_i \equiv M_j$: This identical node N_i is added to NS_P. Two new nodes N_{i+1} and M_{j+1} are selected for next comparison.

Based on analyses above, the algorithm for generating a *nodeset*, the procedure **NodesetGeneration**, is as follows.

```
Procedure NodesetGeneration(Nodeset NS1, Nodeset NS2)
1.    i ← 1; j ← 1; NS;
2.    While((i < NS1.size) ∧ (j < NS2.size))
3.       If(NS1[i].pre-order > NS2[j].pre-order)
4.          If(NS1[i].post-order > NS2[j].post-order) j++;
5.          Else {NS ← NS ∪ NS1[i]; i++;}
6.       Else If(NS1[i].pre-order < NS2[j].pre-order)
7.          If(NS1[i].post-order < NS2[j].post-order) i++;
8.          Else {NS ← NS ∪ NS2[j]; j++;}
9.       Else {NS ← NS ∪ NS1[i]; i++; j++;}
10.   End While
11.   Return NS;
```

It is easy to see that the increasing order of nodes in NS is guaranteed as these nodes are inserted to the end of NS in that order. Therefore, NS is also a *nodeset*.

Lemma 5 (superset equivalence): Given an item I and an itemset $P(I \notin P)$, if the *support* of P is equal to the *support* of $P \cup \{I\}$, the *support* of $A \cup P$ is equal to the *support* of $A \cup P \cup \{I\}$. Here $(A \cap P = \emptyset) \wedge (I \notin A)$.

Proof. As the *support* of itemset P is equal to that of $P \cup \{I\}$, any transaction containing P also contains the item I. Apparently, if a transaction contains $A \cup P$, it must contain P. This means that the numbers of transactions containing $A \cup P$ and $A \cup P \cup \{I\}$ are the same. Therefore, Lemma 5 holds.   ∎

5 Algorithm IFIN⁺

In this section, we introduce the algorithm IFIN⁺ based on its serial version IFIN and the preliminaries introduced in the previous section. For making this article self-contained and easier to refer, algorithm IFIN will be represented here.

5.1 Algorithm IFIN

```
Algorithm 2: IFIN
Input: Stored tree Tree-D, incremental dataset D, ε
Output: Set of frequent k-itemsets L
1.   Create the root node R; L ← ∅;
2.   If(Tree-D <> null) LoadIPPCTree(Tree-D, R, L, transCount);
3.   If(D <> null) BuildIPPCTree(D, R, L);
4.   HasMap<itemset, support> C2 ← ∅;
5.   LOOP:
6.   Ask for a new support threshold ε or exit;
7.   Filter frequent items in L based on ε and add to L1;
8.   If(C2 <> ∅) Goto SKIP;
9.   Scan Each node N in IPPC-Tree with pre order traversal
10.      Iₙ ← N.item-name;
11.      For Each ancestor A of N
12.         Iₐ ← A.item-name;
13.         If(Iₙ < Iₐ) C2.add(IₙIₐ, IₙIₐ.support + N.support);
14.         Else C2.add(IₐIₙ, IₐIₙ.support + N.support);
15.      End For
16.   End Scan
17.   SKIP:
18.   L2' ← L2; L2 ← ∅;
19.   Filter frequent itemsets in C2 based on ε and add to L2;
20.   Scan Each node N in IPPC-Tree with pre order traversal
21.      Iₙ ← N.item-name;
22.      For Each ancestor A of N
23.         Iₐ ← A.item-name;
24.         If(Iₙ < Iₐ) IS ← IₙIₐ;
25.         Else IS ← IₐIₙ;
26.         If((IS ∈ L2)∧(IS ∉ L2')) nodeset_rs.add(N);
27.      End For
28.   End Scan
29.   L ← L ∪ L1; L ← L ∪ L2;
30.   For Each IᵢIⱼ ∈ L2
31.      GenerateFrequentItemsets(IᵢIⱼ, {I|I ∈ L1,Iⱼ < I}, ∅);
32.   Goto LOOP;
```

IFIN algorithm can perform its mining at different support thresholds (lines 5–32) with only one time of constructing the IPPC-Tree (lines 1–4). Lines 9–16 generate the list of candidate 2-itemsets $C2$ as well as their respective *supports*. This task is ignored if the current running session performs for following times of mining with other support thresholds. Lines 20–28 create the corresponding *nodeset* for each frequent 2-itemsets in $L2$. From the second time of mining, just new frequent 2-itemsets' *nodesets* are generated. In lines 30–31, each frequent 2-itemset in $L2$ will be extended by the recursive procedure **GenerateFrequentItemsets** to discover all longer frequent itemsets. This procedure searches a space of itemsets which is demonstrated by a set-enumeration tree [6] constructing from the list of increasingly ordered frequent items $L1$. An example of the search space for the dataset in Table 2 with support threshold $\varepsilon = 0.6$ is visualized in Fig. 4. No exhausted search is performed on the space because of two facts. The first is only sub spaces, whose prefix path are the frequent 2-itemsets, are searched in. The second is the procedure employs two pruning strategies to greatly narrow down the search space. The first strategy is that if P is not a frequent itemset, its supersets are not either; and the second one is the superset equivalence introduced in Lemma 5.

There are three input parameters for procedure **GenerateFrequentItemsets**: (1) *FIS* is a frequent itemset which will be extended. (2) *CI* is a list of candidate items used to expand the *FIS* with one more item. The expansion for *FIS* is based on Definition 2 in Sect. 4. (3) *Parent_FISs* is a set of superset equivalence sets of items generated at the parent of *FIS* in the set-enumeration tree. In that, each superset equivalence set is formed from items which satisfy the condition of Lemma 5 at the parent of *FIS*. In this procedure, lines 1–2 initialize some sets and make a copy *Curr_FISs* of *Parent_FISs*. Lines 3–11 extend the *FIS* with each item I in the item list *CI* that each item will drop in one of the following three cases:

- The extended itemset $FIS \cup \{I\}$ is a frequent itemset: The item I is considered to be a candidate item to expand the search space.
- The supports of *FIS* and $FIS \cup \{I\}$ are equal: The item I satisfies the condition in Lemma 5. Therefore, the item I will be added in to set of superset equivalence items (*eqItems*); and the search in subspaces containing $FIS \cup \{I\}$ in their prefix path does not need to continue because the support of the itemsets $FIS \cup \{I\} \cup \{J_1, \ldots, J_k\}$ $(I \prec J_1 \prec \ldots \prec J_k)$ in these subspaces is equal to the support of itemsets $FIS \cup \{J_1, \ldots, J_k\}$ in other spaces.
- The extended itemset $FIS \cup \{I\}$ is not a frequent itemset: Stop searching all subspaces which contain $FIS \cup \{I\}$ in their prefix path.

Lines 12–21 update the set of superset equivalence sets *Curr_FISs* at the itemset *FIS* in the set-enumeration tree and generate frequent itemsets based on *FIS*, *Curr_FISs* and set of superset equivalence items *eqItems*. Lines 22–25 generate frequent itemsets based on the extended frequent itemsets $FIS \cup \{I\}$ and the updated superset equivalence sets *Curr_FISs*. Finally, the progress of searching subspaces, whose prefix paths are extended frequent itemsets, is continued in lines 26–27.

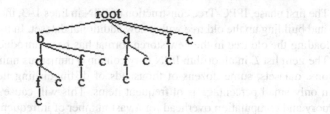

Fig. 4. Set-enumeration tree for example dataset Table 2, support threshold $\varepsilon = 0.6$

```
Procedure GenerateFrequentItemsets(FIS, CI, Parent_FISs)
1.   nextCI ← ∅; eqItems ← ∅; extFISs ← ∅; Curr_FISs ← ∅;
2.   Curr_FISs ← Curr_FISs ∪ Parent_FISs;
3.   For Each item I ∈ CI
4.       IS = (FIS\{FIS.last_item}) ∪ {I};
5.       extIS = FIS ∪ {I};
6.       extIS.nodeset ← NodesetGeneration(FIS.nodeset, IS.nodeset);
7.       If(extIS.support = FIS.support) eqItems.add(I);
8.       Else If(extIS is a frequent itemset){
9.           nextCI.add(I); extFISs.add(extIS); F.add(extIS);
10.      }
11.  End For
12.  If(eqItems <> ∅){
13.      SoS ← set of all subsets of eqItems, excluding ∅;
14.      For Each IS ∈ SoS Do F.add(FIS ∪ IS);
15.      If(Parent_FISs <> ∅){
16.          Production ← {P| P = P1∪P2, P1 ∈ SoS, P2 ∈ Parent_FISs};
17.          For Each IS ∈ Production Do F.add(FIS ∪ IS);
18.          Curr_FISs ← Curr_FISs ∪ Production;
19.      }
20.      Curr_FISs ← Curr_FISs ∪ SoS;
21.  }
22.  If(Curr_FISs <> ∅){
23.      Production ← {P| P = P1∪P2, P1 ∈ extFISs, P2 ∈ Curr_FISs};
24.      F ← F ∪ Production;
25.  }
26.  For Each itemset IS ∈ extFISs
27.      GenerateFrequentItemsets(IS, nextCI, Curr_FISs);
```

5.2   IFIN Parallelization and Improvement

In this subsection, we present the improvement and parallelization to increase efficiency and throughput for IFIN. Algorithm IFIN is divided into four phases and each will be parallelized separately.

The first phase, IPPC-Tree construction of IFIN in lines 1–3, includes loading the old tree and building up the old tree with a new additional dataset. In that, the parallelization for loading the old tree in the new stored format has been introduced in Sect. 3.

The item list $\mathcal{L}$ in algorithm IFIN may contain a numerous number of distinct items in some datasets, some dozens of thousands of distinguishing items, even more; but often only small percentage is of frequent items. This will cause IFIN to waste much memory and computation overhead for a vast number of infrequent itemsets. To reduce this, the item list $\mathcal{L}$ will be replaced by one its subset $\mathcal{L}'$ which is generated by filtering items in $\mathcal{L}$ based on a given lower bound of support thresholds ε' ($\varepsilon' \leq \varepsilon$); and then the items in $\mathcal{L}'$ are sorted in ascending order of their supports before being used in the following phases.

The second phase, frequent 2-itemsets generation, is performed by lines 9–19. Remaining the encoding for each 2-itemset as an ordered string of item names and the set of 2-itemsets $C2$ as a hash map in the parallelized second phase will cause the running time to be not improved, even worse, because of sharing and synchronization between threads when updating 2-itemsets' *supports* in $C2$. To overcome this, each item is encoded as an integer which is its position in the item list $\mathcal{L}'$ ($|\mathcal{L}'| = m$); and instead of a shared hash map $C2$, a $m \times m$ matrix of integers M_t is reserved for t^{th} thread. Two elements $M_t(i, j)$ and $M_t(j, i)$ partially indicate the support of a 2-itemset comprising two items I_i and I_j at positions i and j in $\mathcal{L}'$ respectively. In this phase, the work pool is the built IPPC-Tree, and tasks in the work pool are the built tree's direct sub-trees. When a thread has no longer sub-trees to process, it calculates local supports for 2-itemsets I_iI_j through Eq. (3). After the all threads have completed their work, aggregation and filter operators are performed to achieve the global supports for all 2-itemsets following Eq. (4) and to extract frequent 2-itemsets.

$$Local\_Support_t\left(I_iI_j\right) = M_t(i,j) + M_t(j,i), (i < j) \tag{3}$$

$$Support\left(I_iI_j\right) = \sum\nolimits_t Local\_Support_t\left(I_iI_j\right), (i < j) \tag{4}$$

The third phase, nodesets generation for frequent 2-itemsets, is executed in lines 20–28. The same problems of sharing and synchronization in the second phase happen as threads may concurrently update the same nodeset of a certain frequent 2-itemset. For the purpose of independent execution between threads, nodesets for items need to be generated in advance, and nodesets for frequent 2-itemsets are produced from two nodesets of the two componental items. The work pool is now the list of frequent 2-itemsets, and threads independently retrieve items' nodesets and generate nodesets for frequent 2-itemsets by applying the procedure **NodesetGeneration** in Sect. 4.

The fourth phase, discovering all frequent k-itemsets ($k > 2$) from each frequent 2-itemsets, is executed in lines 30–31. In the set-enumeration tree, the 2-itemsets divide the tree into individual subtrees which are separate subspaces of itemsets. Therefore, threads of searching for longer frequent itemsets in subspaces, which are prefixed by frequent 2-itemsets, are performed independently to each other. The same parallel model as the previous phases, the work pool is a collection of search subspaces corresponding to frequent 2-itemsets. Threads continuously fetch frequent 2-itemsets and start searching the respective subspaces to discover longer frequent itemsets through the procedure **GenerateFrequentItemsets** until the work pool is empty. Based on the above presentation, the algorithm IFIN⁺ is designed as follows.

```
Algorithm 3: IFIN⁺
Input: Stored tree T, incremental dataset D, ε, ε′
Output: Set of frequent k-itemsets L
1.   Create the root node R; L ← ∅;
2.   threadCount ← the number of physical cores in the machine;
3.   If(T <> null) ParallelLoadIPPCTree(T, R, L, threadCount);
4.   If(D <> null) BuildIPPCTree(D, R, L);
5.   L′ ← Filter items in L based on ε′;
6.   Sort items of L′ in ascending order of their supports;
7.   Scan Each node N in IPPC-Tree with pre order traversal
8.      If(L′ contains N.item-name) Nodeset_N.add(N);
9.   LOOP:
10.  Ask for a new support threshold ε or exit;
11.  Filter frequent items in L′ based on ε and add to L1;
12.  If(M₁ <> null) Goto SKIP;
13.  Initialize matrixes M_{t=[1, ThreadCount]} size m x m (m = |L′|);
14.  childIndex ← 0;
15.  For t From 1 To threadCount
16.     Start ItemsetGenThread(R, childIndex, M_t);
17.  Synchronize all threads;
18.  M₁[i,j]_{i,j=[0, m−1]; i<j} = Σ_{t=[1, ThreadCount]} M_t[i,j]_{i,j=[0, m−1]; i<j};
19.  SKIP:
20.  L2′ ← L2; L2 ← ∅;
21.  For each M₁[i, j] ≥ ε * number_of_transactions, (i < j) L2.add(L′[i]L′[j]);
22.  index ← 0;
23.  For t From 1 To threadCount
24.     Start NodesetGenThread(L2\L2′, index);
25.  Synchronize all threads;
26.  L ← L ∪ L1; L ← L ∪ L2; index ← 0;
27.  For t From 1 To threadCount
28.     Start GenerateFrequentItemsetsThread(L2, index);
29.  Synchronize all threads;
30.  Goto LOOP;


GenerateFrequentItemsetsThread(Freq2Itemsets, index)
1.   While(true)
2.      Mutually-exclusive-region {
3.         If(index ≥ Freq2Itemsets.length) break;
4.         I_iI_j = Freq2Itemsets[index]; index++;
5.      }
6.      GenerateFrequentItemsets(I_iI_j, {I|I ∈ L1, I_j < I}, ∅);
7.   End While
```

```
NodesetGenThread(NewFreq2Itemsets, index)
1.   While(true)
2.     Mutually-exclusive-region {
3.         If(index ≥ NewFreq2Itemsets.length) break;
4.         IJ = NewFreq2Itemsets[index]; index++;
5.     }
6.     Nodeset_IJ = NodesetGeneration(Nodeset_I, Nodeset_J);
7.   End While

ItemsetGenThread(R, childIndex, Matrix)
1.   While(true)
2.     Mutually-exclusive-region {
3.         If(childIndex ≥ R.childList.length) break;
4.         subTree = R.childList[childIndex];
5.         childIndex++;
6.     }
7.   Scan Each node N of subTree with pre order traversal
8.       i = mapToIndex(N.item-name);
9.     If (i is invalid) continue;
10.    For Each ancestor A of N
11.        j = mapToIndex(A.item-name);
12.      If (j is invalid) continue;
13.      Matrix[i,j] = Matrix[i,j] + N.support;
14.      End For
15.    End Scan
16.  End While
17.  For i From 0 To Matrix.with-1
18.    For j From i+1 To Matrix.with-1
19.        Matrix[i,j] = Matrix[i,j] + Matrix[j,i];
```

6 Experiments

All experiments were conducted on a 1.86 GHz Intel Core (MT) i3-4030U processor, and 4 GB memory computer with Window 8.1 operating system. To evaluate the algorithms, we used the Market-Basket Synthetic Data Generator [7] based on the IBM Quest to generate a synthetic dataset, and a real dataset named Kosarak [20], online news portal click-stream data. The datasets' properties are shown in Table 4.

The algorithm IFIN⁺ was compared with its original version IFIN, two state-of-the-art algorithms FIN and PrePost⁺, and the well-known one FP-Growth. All the five algorithms were implemented in Java. Experimental values of running time and used memory are the average values from three corresponding individual ones. In our previous works [18, 19], to guarantee available memory of 2 GB used for Java Heap, we set value "-Xmx2G" for _JAVA_OPTIONS, a Windows environment variable. However, we realize that this causes the Java Garbage Collector (GC) to run many times of full memory collection; and consequently, the total running time includes a significant percentage, approximate 45%, for the garbage collection. To avoid this in our current work, we set the value "-Xms2G-Xmx2G" for _JAVA_OPTIONS instead, and result in the running time for garbage collection is reduced to 6% which reflects more exactly the algorithms' performance.

Table 4. The datasets' properties

	No. of transactions	Max length	Average length	No. of total distinct items	No. of frequent items at thresholds		
					0.001	0.002	0.006
Synthetic dataset	1200000	32	10	932	843	774	525
Kosarak	990002	2498	8.1	41270	1260	568	116

For emulating scenarios of incremental mining, the synthetic dataset was divided into six equal parts, 200 thousand transactions for each one, and so on for Kosarak dataset with five parts in which the last one contains just 190002 transactions. The experiments start mining on the first part and then part by part from the second one is accumulated and mined. Both IFIN and IFIN⁺ can perform following three scenarios:

- **S1** (Incremental in Different Sessions): An IPPC-Tree corresponding with a dataset had been constructed, mined and stored in a running session. In the following sessions, the old tree is loaded and then built up with a new additional dataset.
- **S2** (Incremental in the Same Session): An IPPC-Tree corresponding with a dataset has been constructed and mined, and then it is built up with a new additional dataset in the same session.
- **S3** (Just Loading Tree): A stored IPPC-Tree in a previous session is loaded and mined in the following sessions.

Each execution scenario can be performed with different support thresholds in the same running session. The processor in our computer possesses two physical computational units, and we found that the performance achieved its best with two threads in parallel version IFIN⁺. The experiments will be presented in three parts: comparisons between IFIN⁺ and IFIN, and IFIN⁺ against FP-Growth, FIN, and PrePost⁺ on each of the two datasets.

6.1 IFIN⁺ Against IFIN

In this subsection, we present comparisons on the running time of the four partial processing phases between IFIN⁺ and IFIN on the synthetic and Kosarak datasets.

Table 5 reports the running time in seconds of the execution phases for IFIN⁺ and IFIN on the synthetic dataset in increasing sizes from 200k to 1200k transactions. In the

Tree Loading phase, the IFIN$^+$'s execution is speeded up by more than $3\times$ compared to that of IFIN. More contrast, an approximate $10\times$ speedup for IFIN$^+$ is achieved in the Frequent 2-itemset Generation phase. In the third phase, the IFIN$^+$'s running time is reduced to two-thirds of the running time of IFIN; and in the final phase, the performance of IFIN$^+$ is also improved compared to that of IFIN, approximate a quarter of the running time of IFIN is reduced for IFIN$^+$.

Table 5. Running time between IFIN$^+$ and IFIN on the synthetic dataset at different sizes

	200k	400k	600k	800k	1000k	1200k
Running time in IPPC-Tree loading phase						
IFIN	1.4 s	2.4 s	3.6 s	4.7 s	5.8 s	7.4 s
IFIN$^+$	0.4 s	0.6 s	1.1 s	1.4 s	1.9 s	2.1 s
Running time in Frequent 2-itemset Generation phase $(\varepsilon = 0.001)$						
IFIN	3.1 s	5.4 s	8	10.1 s	11.8 s	13.8 s
IFIN$^+$	0.4 s	0.5 s	0.8 s	0.9 s	1 s	1.4 s
Running time in Nodeset Generation phase $(\varepsilon = 0.001)$						
IFIN	2.2 s	4.4 s	6 s	7.2 s	8.6 s	10.5 s
IFIN$^+$	1.4 s	2.4 s	3.7 s	4.3 s	6 s	6.3 s
Running time in Discover Frequent k-itemsets phase $(\varepsilon = 0.001)$						
IFIN	1.9 s	3 s	4 s	5.8 s	6.9 s	8.2 s
IFIN$^+$	1.6 s	2.3 s	3.1 s	4.6 s	5.0 s	6.1 s

Table 6. Total running time[a] of the three versions on the synthetic dataset

Scenario S1, $\varepsilon = 0.001$	IFIN [18]	IFIN$^+$ [19]	IFIN$^+$ ([19] applies the new stored format for IPPC tree, and the parallelization for Discover Frequent k-itemsets phase)
200k	10.3 s	6.1 s	5.7 s
400k	17.7 s	10 s	8.5 s
600k	24.3 s	14 s	11.6 s
800k	33.2	18.6	14.8 s
1000k	37.6 s	22.5 s	17 s
1200k	43.8 s	26.6 s	20 s

[a]Beside the running time of the four presented phases, the total running time includes the time for building up the loaded tree with an additional dataset, generating nodesets for items, writing result, and parts of the Java GC time.

Table 6 summarizes the total running time on the synthetic dataset (with scenario S1 and support threshold = 0.001) of the three versions IFIN [18], IFIN$^+$ [19], and IFIN$^+$ [19] applying the new stored format for IPPC tree and the parallelization for Discover Frequent k-itemsets phase. As we can see, the performance of the current version of IFIN$^+$ is highest and more than twice higher than that of IFIN. For the used memory, there are no differences among the three versions.

Table 7. Running time between IFIN$^+$ and IFIN on Kosarak datasets at different sizes

	200k	400k	600k	800k	1000k
Running time in IPPC-Tree loading phase					
IFIN	1.3 s	2.5 s	3.7 s	4.8 s	6.2 s
IFIN$^+$	0.5 s	0.7 s	1.2 s	1.5 s	2.1 s
Running time in Frequent 2-itemset Generation phase ($\varepsilon = 0.002$)					
IFIN	6.6 s	13.2 s	18.2 s	24 s	28.5 s
IFIN$^+$	0.6 s	1.1 s	1.7 s	2.3 s	2.5 s
Running time in Nodeset Generation phase ($\varepsilon = 0.002$)					
IFIN	4.2 s	7.5 s	10.3 s	13.2 s	15.5 s
IFIN$^+$	0.2 s	0.3 s	0.5 s	0.6 s	0.8 s
Running time in Discover Frequent k-itemsets phase ($\varepsilon = 0.002$)					
IFIN	0.9 s	1.8 s	3.5 s	5.6 s	7.1 s
IFIN$^+$	1 s	2.4 s	3.4 s	4.8 s	6.7 s

Table 7 shows the running time of the four processing phases for IFIN$^+$ and IFIN on Kosarak dataset in increasing sizes from 200k to 990002 transactions. Like the synthetic dataset, the same ratios of performance improvement for IFIN$^+$, approximate speedups of 3× and 10× are respectively achieved in the first and the second phases for Kosarak dataset. While there is no much running time reduction in the final phase, a very sharp contrast of performance is revealed in the third phase between IFIN$^+$ and IFIN, approximate 20× faster for IFIN$^+$ in most data accumulation steps. The third processing phase makes a great difference between two datasets, 1.66× vs. 20× speedup for the synthetic dataset and the real one Kosarak respectively. The reason for this is as follows.

As introduced in Sect. 5, the third phase of IFIN is based on traversing on the IPPC-Tree while the third phase of IFIN$^+$ manipulates on the nodesets of items in $\mathcal{L}'$ to generate nodesets for all frequent 2-itemsets. The Kosarak dataset is much sparser than the synthetic dataset (refer the Table 4) so the IPPC-Tree corresponding to Kosarak dataset tends to be larger than the tree of the synthetic dataset. The number of frequent 2-itemsets of Kosarak dataset, 2469 in average, is much lesser than 9411 frequent 2-itemsets of the synthetic dataset. Moreover, the nodesets' average length in Kosarak dataset is also shorter than that in the synthetic dataset.

Table 8. Total running time of the three versions for Kosarak dataset

Scenario S1, $\varepsilon = 0.002$	IFIN [18]	IFIN⁺ [19]	IFIN⁺ ([19] applies the new stored format for IPPC tree, and the parallelization for Discover Frequent k-itemsets phase)
200k	21.6 s	10 s	9.2 s
400k	36.4 s	15 s	14.7 s
600k	47.7 s	18.6 s	17.4 s
800k	61.2 s	24.2 s	22.1 s
1000k	72.1 s	28.3 s	25.3 s

Similar to Tables 6 and 8 reports the total running time of the three versions on Kosarak dataset (with scenario S1 and support threshold = 0.002). The current IFIN⁺ is the most efficient algorithm. Its running time approximates one third that of IFIN, and there are no differences in used memory among the three versions.

In an overview of experiments on both datasets, the current IFIN⁺'s execution time in phases is reduced substantially compared to that of its original version IFIN. The larger the number of transactions is, the more the running time is saved for IFIN⁺. The reason for such remarkable performance enhancement is that the efficient parallelization is synergized with improvement in data representation.

6.2   Experiments on the Synthetic Dataset

In this subsection, we benchmark the running time and peak used memory of IFIN⁺ against that of the three algorithms FP-Growth, FIN and PrePost⁺ on the synthetic dataset. In that, the algorithm IFIN⁺ experiments with all its possible execution scenarios S1, S2, and S3 as referred.

Figures 5 and 6 sequentially demonstrate the running time and peak memory of the algorithms in steps of data accumulation at the support threshold $\varepsilon = 0.1\%$. For all algorithms, both running time and peak memory increase linearly when the dataset is accumulated. While FP-Growth is the slowest algorithm, it uses memory more efficient than FIN and PrePost⁺. Algorithm PrePost⁺ consumes the most memory, but it runs faster than FP-Growth and FIN. Algorithm IFIN⁺ is the most efficient for both running time and consumed memory. Follow the increasing of the dataset size, while the advantage in used memory of IFIN⁺ remains stable; the execution time of IFIN⁺ becomes more dominant, lesser than a haft, compared to the remaining algorithms'. Among the three execution scenarios of IFIN⁺, the running time of S2 and S3 is almost the same and better than S1's but not much difference. Beside the high performance of mining phases in algorithm IFIN⁺, one more reason can be found out in Table 9 which benchmarks the algorithms' tree construction time.

Note that the IPPC-Tree construction does not depend on support threshold, but the other three trees. The trees of FIN and PrePost⁺ are almost the same, so their running time for tree building is nearly equal. The tree construction of IFIN⁺ achieves the best performance, much better than the three algorithms'. Especially in scenario S3, the time ratios are 1:7 and 1:6 compared to FP-Growth's and PrePost⁺'s respectively. At the same dataset size, building tree in S3 is faster than that in S1, approximate 2.3 s in average; since the execution scenario S1 must build up the loaded tree with a new additional dataset of 200k transactions. This also reveals that constructing an IPPC-Tree by loading its stored data is much efficient than building that tree from the corresponding dataset of transactions.

Table 9. Tree construction time of the algorithms for the synthetic dataset

	200k	400k	600k	800k	1000k	1200k
IFIN⁺ (S1)	2.1	3.0	3.2	4.1	4.3	5.1
IFIN⁺ (S3)	0.4	0.6	1.1	1.4	1.9	2.1
FP-Growth ($\varepsilon = 0.001$)	3.4	5.7	8.7	10	12.9	16.2
Fin/PrePost $+ (\varepsilon = 0.001)$	2.5	4.5	6.7	9.7	11.9	14.7

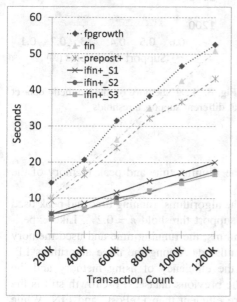

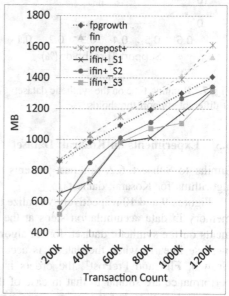

Fig. 5. Running time on the incremental synthetic datasets

Fig. 6. Peak memory on the incremental synthetic datasets

In Figs. 7 and 8, the running time and peak used memory are visualized for the algorithms mining on the synthetic dataset of 1.2 million transactions with different support thresholds ε. Start at $\varepsilon = 0.6\%$, IFIN⁺ can perform one of two scenarios S1 or S3 that their two running time values are shown in Fig. 7. For other ε values, IFIN⁺ just run its mining tasks since the built tree is completely reused. Furthermore, only a portion of its mining is performed. Consequently, with following values of $\varepsilon < 0.6\%$, the running time of IFIN⁺ takes an overwhelming dominance against that of the three algorithms. The consumed memory of IFIN⁺ is lesser than the rest algorithms'. The algorithm FP-Growth uses memory more efficient than the two algorithms FIN and PrePost⁺. However, its running time is considerably longer than that of FIN and Pre-Post⁺. Algorithm PrePost⁺ run faster than FIN and FP-Growth, but it uses the most memory.

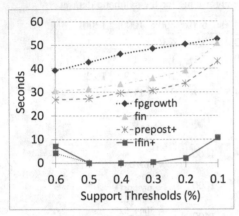
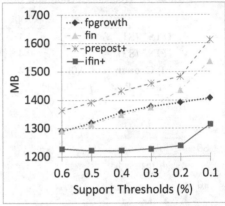

Fig. 7. Running time on the synthetic dataset at different support thresholds

Fig. 8. Peak memory on the synthetic dataset at different support thresholds

6.3 Experiments on Kosarak Dataset

Similar to Subsect. 6.2, this part presents the running time and peak memory of the algorithms for Kosarak dataset.

Figures 9 and 10 respectively visualize the algorithms' running time and peak used memory in data accumulation steps at the support threshold $\varepsilon = 0.2\%$. Like experiments on the synthetic dataset, for all algorithms, the running time and used memory increase linearly when the dataset is accumulated. Among the three algorithms FP-Growth, FIN and PrePost⁺, the orders in the efficiency of using memory and the performance are similar to that in case of the previous dataset. FP-Growth still is the slowest algorithm, but it uses memory more efficient than PrePost⁺ and FIN. While PrePost⁺ runs considerably faster than FP-Growth and FIN, it becomes to consume more memory than FP-Growth and FIN follow the increasing of dataset size.

However, between IFIN⁺ and the rest three algorithms, there are some changes in running time and used memory. In Fig. 10, at the dataset of 200k transactions, IFIN⁺ uses memory better than the others. However, its used memory increases faster than the other algorithms' and is the most for larger sizes, approximate the PrePost⁺'s memory at full size of Kosarak dataset in scenarios S1 and S3. In contrast to this, the corresponding experiments on the synthetic dataset in Fig. 6 show that IFIN⁺ consumes the least memory for all sizes and execution scenarios.

Table 10. Tree construction time of the algorithms for Kosarak dataset

	200k	400k	600k	800k	1000k
IFIN⁺ (S1)	7.2 s	10.4 s	11.4 s	13.9 s	14.7 s
IFIN⁺ (S3)	0.5 s	0.7 s	1.2 s	1.4 s	2.1 s
FP-Growth ($\varepsilon = 0.002$)	3.3 s	6.2 s	9.2 s	12.6 s	15.3 s
Fin/PrePost + ($\varepsilon = 0.002$)	2.1 s	3.7 s	5.1 s	6.6 s	7.5 s

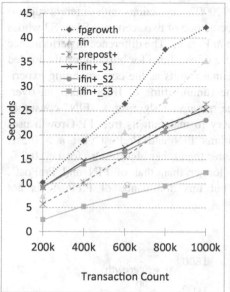

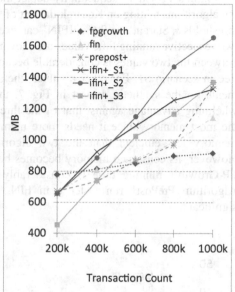

Fig. 9. Running time on the incremental Kosarak datasets

Fig. 10. Peak memory on the incremental Kosarak datasets

As we knew that the IPPC-Tree of IFIN⁺ is a compact structure of all items in a dataset, but the trees of the rest algorithms depend on the support threshold and contain only frequent items in a dataset. Looking into Table 4 for a clear reason, the synthetic dataset comprises a considerable percentage of frequent items, [90% − 56%] for the support threshold $\varepsilon \in [0.001 − 0.006]$; but just a very small quantity, [1.38% − 0.28%] for $\varepsilon \in [0.002 − 0.006]$, is for frequent items in Kosarak dataset. Therefore, in the case of Kosarak dataset, the used memory to maintain the IPPC-Tree of IFIN⁺ is much larger than that of the trees of FP-Growth, FIN and PrePost⁺; while the affection of this disadvantage to IFIN⁺ is not considerable in the synthetic dataset case.

Beside the memory, computational overhead to construct the IPPC-Tree is also more than that of the other trees in the case of Kosarak dataset. Table 10 reports the running time for building the trees of algorithms on Kosarak dataset. Building up the IPPC-Trees takes most of the total time of the IFIN⁺'s tree construction in scenario S1, average 12.5 s for each accumulated dataset of 200k transactions. However, the tree construction in execution scenario S3, just by loading the built tree, takes the least time and once again asserts its very high performance. The tree constructions of FIN and PrePost+ on this dataset are very efficient and take only 7.5 s for the full size of Kosarak dataset. The above facts have explained the trend of the algorithms' running time in the Fig. 9. Execution scenario S3 of IFIN+ runs fastest for all steps of data accumulation while the running time of scenarios S1 and S2 becomes better than the all remaining algorithms' until the step of 800k transactions.

Figures 11 and 12 sequentially depict the running time and peak used memory of the algorithms mining on Kosarak dataset of 990002 transactions with different support thresholds ε. Start at $\varepsilon = 0.6\%$, IFIN+ can execute one of two scenarios S1 or S3 whose two respective running time values are shown in Fig. 11. The difference in performance between the two values is considerable because of overhead for building up the loaded tree in scenario S1. For other ε values, the same results as the corresponding experiments on the synthetic dataset in Fig. 7, the running time of IFIN+ takes an overwhelming advantage against that of the three remaining algorithms. IFIN+ consumes the most memory since it needs more memory to maintain its tree. FP-Growth uses memory less efficient than the two algorithms FIN and PrePost+ for $\varepsilon > 0.3\%$. However, its consumed memory becomes better than other algorithms' for $\varepsilon \le 0.3\%$. FP-Growth's running time is considerably longer than that of FIN and PrePost+. Algorithm PrePost+ run faster than FIN, but this dominance of PrePost+ is not significant.

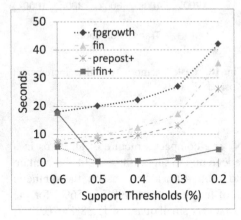
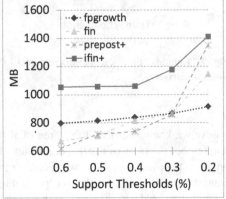

Fig. 11. Running time on Kosarak dataset at different support thresholds

Fig. 12. Peak memory on Kosarak dataset at different support thresholds

7 Conclusions

In this paper, we proposed a solution, IFIN⁺, for parallelizing the frequent itemsets mining algorithm IFIN. Most portions of the serial version were changed in means which increases the efficiency and computational independence for convenience in designing parallel computation with the load balance model, Work-Pool. Hence, computational throughput and efficiency are increased significantly compared to its serial version.

We conducted extensive experiments on a synthetic and a real dataset Kosarak to evaluate the efficiency of both running time and peak consumed memory of IFIN⁺ against the well-known algorithm FP-Growth and two state-of-the-art ones FIN and PrePost⁺. The experiments showed that for a dataset which a large enough percentage of its items is used to construct the corresponding trees of FP-Growth, FIN and Pre-Post⁺, the efficiency of IFIN⁺ on execution time and used memory takes clear dominance over these algorithms'. Otherwise, with a dataset having a small percentage of such items in a huge number of distinct items, IFIN⁺ reduces its advantage since it uses more computation and memory overhead than the three remaining algorithms to construct and retain its IPPC-Tree. However, IFIN⁺ is compensated by the incremental abilities to construct its tree structure and mining that is very helpful to deal with the high-velocity property of Big Data and the data mining practices which often try with different threshold values. This allows IFIN⁺ to save some computational overhead when support threshold changes and not to waste time to rebuild the IPPC-Tree when new data is accumulated. The IPPC-Tree is a compact structure of all items in a dataset. Therefore, the tree can be completely reused to mine with different support thresholds in the same session of the tree building; or in following sessions, the built tree is loaded very efficiently, much faster than constructing the tree from the original dataset in all the four algorithms.

The aim of shared-memory based parallelized algorithm IFIN⁺ is to increase the throughput for its serial version IFIN by utilizing as much as possible the computational power of commodity multi-cores processors. In fact, to deal with the running time problem in Big Data, it is just a minor solution. For a major and much preferred one, a parallelization solution for IFIN⁺ on the distributed environment will be proposed to better confront with the running time and memory scalability problems of Big Data.

References

1. Agrawal, R., Srikant, R.: Fast algorithms for mining association rules. In: Proceedings of 20th International Conference on VLDB, pp. 487–499 (1994)
2. Han, J., Pei, J., Yin, Y.: Mining frequent itemsets without candidate generation. ACM SIGMOD Rec. **29**(2), 1–12 (2000)
3. Cheung, W., Zaïane O.R.: Incremental mining of frequent patterns without candidate generation or support constraint. In: Proceedings of the 7th International Database Engineering and Applications Symposium, pp. 111–116. IEEE (2003)

4. Deng, Z.-H., Lv, S.-L.: Fast mining frequent itemsets using nodesets. Expert Syst. Appl. **41** (10), 4505–4512 (2014)
5. Deng, Z.-H., Lv, S.-L.: PrePost⁺: an efficient N-lists-based algorithm for mining frequent itemsets via children-parent equivalence pruning. Expert Syst. Appl. **42**(13), 5424–5432 (2015)
6. Rymon, R.: Search through systematic set enumeration. In: Proceedings of the 1st International Conference on Principles of Knowledge Representation and Reasoning, pp. 539–550 (1992)
7. Market-Basket Synthetic Data Generator. https://synthdatagen.codeplex.com/
8. Savasere, A., Omiecinski, E., Navathe, S.: An efficient algorithm for mining association rules in large databases. In: VLDB, pp. 432–443 (1995)
9. Perego, R., Orlando, S., Palmerini, P.: Enhancing the apriori algorithm for frequent set counting. In: International Conference on Data Warehousing and Knowledge Discovery, pp. 71–82 (2001)
10. Park, J.S., Chen, M.S., Yu, P.S.: Using a hash-based method with transaction trimming and database scan reduction for mining association rules. IEEE Trans. Knowl. Data Eng. **9**(5), 813–825 (1997)
11. Zaki, M.J.: Scalable algorithms for association mining. IEEE Trans. Knowl. Data Eng. **12**(3), 372–390 (2000)
12. Grahne, G., Zhu, J.: Fast algorithms for frequent itemset mining using FP-Trees. Trans. Knowl. Data Eng. **17**(10), 1347–1362 (2005)
13. Liu, G., Lu, H., Lou, W., Xu, Y., Yu, J.X.: Efficient mining of frequent itemsets using ascending frequency ordered prefix-tree. DMKD J. **9**(3), 249–274 (2004)
14. Shenoy, P., Haritsa, J.R., Sudarshan, S.: Turbo-charging vertical mining of large databases. In: 2000 SIGMOD, pp. 22–33 (2000)
15. Zaki, M.J., Gouda, K.: Fast vertical mining using diffsets. In: 9th SIGKDD, pp. 326–335 (2003)
16. Liu, J., Wu, Y., Zhou, Q., Fung, B.C.M., Chen, F., Yu, B.: Parallel eclat for opportunistic mining of frequent itemsets. In: Chen, Q., Hameurlain, A., Toumani, F., Wagner, R., Decker, H. (eds.) DEXA 2015. LNCS, vol. 9261, pp. 401–415. Springer, Cham (2015). https://doi.org/10.1007/978-3-319-22849-5_27
17. Yun, U., Lee, G.: Incremental mining of weighted maximal frequent itemsets from dynamic databases. Expert Syst. Appl. **54**, 304–327 (2016)
18. Huynh, V.Q.P., Küng, J., Dang, T.K.: Incremental frequent itemsets mining with IPPC tree. In: Benslimane, D., Damiani, E., Grosky, W.I., Hameurlain, A., Sheth, A., Wagner, R.R. (eds.) DEXA 2017. LNCS, vol. 10438, pp. 463–477. Springer, Cham (2017). https://doi.org/10.1007/978-3-319-64468-4_35
19. Huynh, V.Q.P., Küng, J., Jäger, M., Dang, T.K.: IFIN⁺: a parallel incremental frequent itemsets mining in shared-memory environment. In: Dang, T.K., Wagner, R., Küng, J., Thoai, N., Takizawa, M., Neuhold, E.J. (eds.) FDSE 2017. LNCS, vol. 10646, pp. 121–138. Springer, Cham (2017). https://doi.org/10.1007/978-3-319-70004-5_9
20. Frequent Itemset Mining Dataset Repository: Kosarak, Online News Portal Click-Stream Data. http://fimi.ua.ac.be/data/kosarak.dat.gz

Automated Security Analysis of Authorization Policies with Contextual Information

Khai Kim Quoc Dinh and Anh Truong$^{(\boxtimes)}$

Ho Chi Minh City University of Technology, VNU-HCM,
Ho Chi Minh City, Vietnam
anhtt@hcmut.edu.vn

Abstract. Role-Based Access Control (RBAC) has made great attention in the security community and is widely deployed in the enterprise as a major tool to manage security and restrict system access to unauthorized users. As the RBAC model evolves to meet enterprise requirements, the RBAC policies will become complex and need to be managed by multiple collaborative administrators. The collaborative administrator may interact unintendedly with the policies, creates the undesired effect to the security requirements of the enterprise. Consequently, researchers have studied various safety analyzing techniques that are useful to prevent such issues in RBAC, especially with the Administrative Role-Based Access Control (ARBAC97). For critical applications, several extensions of RBAC, such as Spatial-Temporal Role-Based Access Control (STRBAC), are being adopted in recent years to enhance the security of an application on authorization with contextual information such as time and space. The features, which proposed in STRBAC for collaborative administrators, may interact in subtle ways that violate the original security requirements. However, the analysis of it has not been considered in the literature.

In this research, we consider the security analysis technique for the extension of STRBAC, named Administrative STRBAC (ASTRBAC), and illustrate the safety analysis technique to detect and report the violation of the security requirements. This technique leverages First-Order Logic and Symbolic Model Checking (SMT) by translating the policies to decidable reachability problems, which are essential to understand the security policies and inform policies designer using this model to take appropriate actions. Our extensive experimental evaluation demonstrates the correctness of our proposed solutions in practice, which supports finite ASTRBAC policies analysis without prior knowledge about the number of users in the system.

Keywords: Computer security · Security analysis · Access control ·
Role-Based Access Control · Spatial-Temporal Role-Based Access Control

1 Introduction

Role-based access control (RBAC) [4, 5] is gaining more attention in protecting enterprise data against malicious users and criminals. Access control [1] is essential to mediate every request to enterprise resources and decide whether to approve or reject each request. Manipulating these requests requires access control policies to define

© Springer-Verlag GmbH Germany, part of Springer Nature 2019
A. Hameurlain et al. (Eds): TLDKS XLI, LNCS 11390, pp. 107–139, 2019.
https://doi.org/10.1007/978-3-662-58808-6_5

higher level rules to regulate and control who and what kind of permissions can access specific resources. Although many access control models have been developed like Discretionary Access Control (DAC) [2] or Mandatory Access Control (MAC) [3], only RBAC achieves awareness against users and vendors due to its benefits for organizations which separates responsibilities in a system where multiple roles are fulfilled.

In RBAC, the access permissions are associated with roles, and users must be made members of appropriate roles to grant access. Using Principle of Least Privilege and Separation of Duties, no one can have discretionary access to enterprise resources to perform malicious activities since no standalone individual has all permissions needed for an important operation. RBAC simplifies management of authorization and flexible in specifying and enforcing policies which can be added or removed for a particular role as needed. Research works [5, 6] have been dedicated to expanding RBAC model to support enterprises in management policies where the number of users and administrators keep increasing. Administrative Role-Based Access Control (ARBAC) [7] is proposed for changing management in policies by administrators and decentralized policies administration. As the enterprises grow, the policies may be modified frequently by adding or removing some tuples in the policies when the employee changes their job or gets promoted. The changing process might be done by many administrators, each of whom can make small changes to parts of the policies that, at first look, seems harmless. However, when applying, the combination of all of these changes may lead to an unsafe state which violates security properties of the policies. Therefore, it is necessary to have a solid change management solution which checks for vulnerabilities and violations in security before applying those changes to the system. This vulnerable in Administrative Role-Based Access Control (ARBAC97) [10] have been intensively done [17].

Over the last few year, several extensions of RBAC such as Spatial-Temporal Role-Based Access Control (STRBAC) [8] are being focused to enhance the security of an application on authorization with contextual information such as time and space; however, there isn't much focus on the security analysis of the Administrative model of STRBAC (ASTRBAC) [9]. In order to overcome these shortcomings, in this research, we propose a security analysis technique for ASTRBAC based on First-Order Logic and Symbolic Model Checking [18]. The main idea is to adapt First-Order Logic and Symbolic Model Checking to translate the security analysis problem of ASTRBAC policies to decidable reachability problem where total users and roles are finite but the exact number is not known in order to mechanize the analysis. Based on the model checking proposed in [16], we create a framework to help security officers aware of the existence of vulnerabilities in the policies before applying those policies to production systems. This model can also return the group of actions which cause the vulnerability to help security officers in detecting and modifying security policies easier according to their needs and keep compliance with security requirements of the organization.

The paper is organized as follows: Sect. 2 briefly introduces RBAC, its administrative model and STRBAC model. Section 3 presents our automated analysis technique for STRBAC policies. Our extensive experiment is illustrated in Sect. 4. Finally, Sect. 5 concludes the paper.

2 Role-Based Access Control, Spatial-Temporal Role-Based Access Control, and Administration

2.1 RBAC

In 1992, Ferraiolo and Kuhn firstly proposed Role-based Access Control model, which was standardized by National Institute of Standards and Technology (NIST) [6]. Its main idea is that permissions are associated with roles and users must be assigned to appropriate roles to gain those permissions.

RBAC has been considered as an alternative to the well-known tradition access controls such as DAC and MAC. In general, the RBAC policies are a tuple *(U, R, P, UA, PA)* which consists of a set *U* of users, a set *R* of Roles, a set *P* of Permissions, a User-Role Assignment relation $UA \subseteq U \times R$, a Role-Permission Assignment relation $PA \subseteq R \times P$, and for simplicity, we ignore the role hierarchy (see [15, 29] for more details).

As stated in RBAC, a user *u* is a member of a role *r* if *(u, r)* $\in UA$; a role *r* is assigned permission *p* if *(p, r)* $\in PA$. Thus, a user is granted to permission *p* if and only if there exists a role $r \in R$ such that *(p, r)* $\in PA$ and *u* are the member of *r*. The *UA* relations in RBAC keep changing according to the growth and reduction of human resources in an organization while the *PA* will be less likely to change because of the fact that the change of this part means there is a change in organization structure and this may impact the entire system.

RBAC model has the advantage of being simple, efficient, and convenient for management, which is suitable for many applications in reality. However, the RBAC model still has many limitations such as not suitable for applications with security resources that are unknown; not appropriate for complex access control rules when access control is based not only on roles but also on many other contextual elements.

2.2 Requirement for RBAC Policies Change Management

Typically, RBAC policies consist of two parts *UA* and *PA*. A user *u* can use a permission *p* of the system if and only if user *u* is granted a role *r* and a permission *p* has been assigned to that role *r*.

User	Role
u1	Trainer
u2	Intern

User Assignment (UA)

Role	Permission
Tester	WriteObject
InternAssistant	ReadObject
Trainer	WriteObject
Intern	AppendObject

Permission Assignment (PA)

Fig. 1. RBAC

In Fig. 1, the user *u1* is assigned the Trainer role, the user *u2* is assigned the Intern role in a company system (table *UA*). Therefore, *u1* is granted write access to the system, $u2$ has the permission to insert into the system because the *Trainer* has been granted write access to the system, the *Intern* role is granted permission to insert into the system (table *PA* of policies). We also notice that $u2$ does not write to the system because the intern role is not granted access to that system.

In reality, the policies will keep changing, for example, when employees are promoted or leaving, which makes the policies need to be updated by deleting, modifying, or adding new rules to reflect the change of organization. The role assignment in *UA* is frequently changed as new employees entering or leaving the job or changing their roles that led to the *UA* being edited while assigning permission in *PA* is less likely to change, since changing the *PA* will change the structure of the organization. Therefore, in our research, we will focus on the change management in the *UA* of the policies.

2.3   Administrative Role-Based Access Control (ARBAC)

The ARBAC is the most accepted administrative framework to control how RBAC policies might change through administrative actions by assigning or revoking user memberships into roles (URA model, a sub-model of ARBAC [7]).

The ARBAC model consists of three main components: User to Role Assignment (URA97), Role to Permission Assignment, and Role to Role Assignment (RRA97). URA97 and PRA97 manipulate changes on *UA* and *PA* relationships by assigning rules or revoking rules in administrator actions: *can_assign* and *can_revoke*, respectively. RRA97 controls the change in the Role Hierarchies by modifying rules in administrator actions can_modify. However, in our research, we focus on the changes in RBAC policies by assignment rules and revoking rules.

In the URA97 model, administrators can only update the relations in *UA* using the defined administrative actions while the relations in *PA* keeps constant. The first administrative action is to assign users to roles and is defined using ternary relation *can_assign(A, C, r)* where *A* (called Administrative) and *C* (called Simple) are preconditions and *r* is a role (called target role). The second administrative action is to revoke users from roles and is defined using binary relation *can_revoke(A,r)* where *A* (called Administrative) is a pre-condition and *r* is a role (called target role). A pre-condition *C* is defined as a finite set of signed role, which expressed using $+r$ or $-r$ for $r \in R$. We can say that a user $u \in U$ satisfied a pre-condition *C* (written as $u \vDash C$) if for each $c \in C$, u is the member of $r \in R$ when c is $+r$ and u is not a member of $r \in R$ when c is $-r$.

For example, consider a *can_assign ({+Manager}, {+Trainer, −Intern}, Group-Lead)*, this rule allows administrators with role *Manager* to assign any users, who have been assigned the role *Trainer* and haven't been assigned the role Intern, to a new role *GroupLead* in the company training program. Consider the *can_revoke ({+Manager}, Intern)*, this rule allows administrators with role *Manager* to revoke any users who have been assigned the role Intern in the company training program.

2.4   User-Role Reachability Problem

While URA97 restrictions can limit the administrative actions, research [11] has found that the change to RBAC policies by one administrator may interact in unintended or malicious ways with other administrator's actions. This problem is well-known as the safety problem (also called the reachability problem), which the effects of these interactions may lead to an unintended role assignment to an untrusted user, and let that user have the ability to view or stole sensitive information or resources.

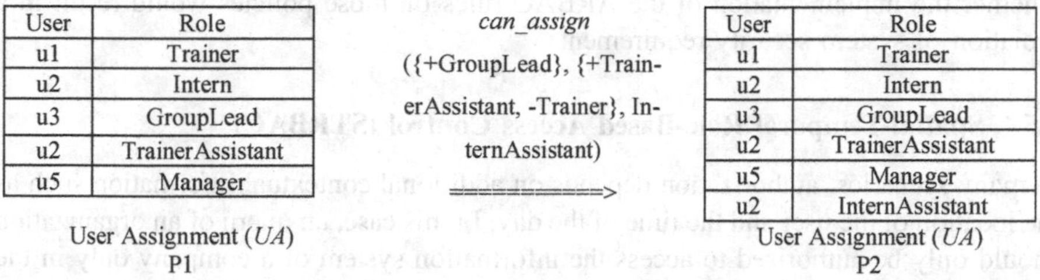

User	Role
u1	Trainer
u2	Intern
u3	GroupLead
u2	TrainerAssistant
u5	Manager

User Assignment (*UA*)
P1

can_assign
({+GroupLead}, {+Train-
erAssistant, -Trainer}, InternAssistant)
=======>

User	Role
u1	Trainer
u2	Intern
u3	GroupLead
u2	TrainerAssistant
u5	Manager
u2	InternAssistant

User Assignment (*UA*)
P2

Fig. 2. Policies P1 is changed to P2

Consider the system that has policies *P1* as shown in previous Fig. 1 and a set of ARBAC rules as follows:

Rule 1: *can_assign({+GroupLead}, {+TrainerAssistant, −Trainer}, InternAssistant);*
Rule 2: *can_assign({+Manager}, {+InternAssistant}, Tester)*

The system requires that a user with the role *Intern* is not allowed to grant the role *Tester*. This requirement is often referred as the security requirements of the system, also known as the security attributes. In Fig. 2, Rule 1 can be executed by *GroupLead u3* to assign the user *u2* the role *InternAssistant* as mentioned in the previous example. After executing this rule, policies *P1* will be modified as shown in Fig. 3. Next, using Rule 2 on policies P2 *can_assign({+Manager}, {+InternAssistant}, Tester).*

- *u5* satisfies the administrative condition *{+manager}* because *(u5, manager)* belongs to *p2*
- *u2* satisfies the user condition *{+internassistant}* because *(u2, internassistant)* belongs to *p2*

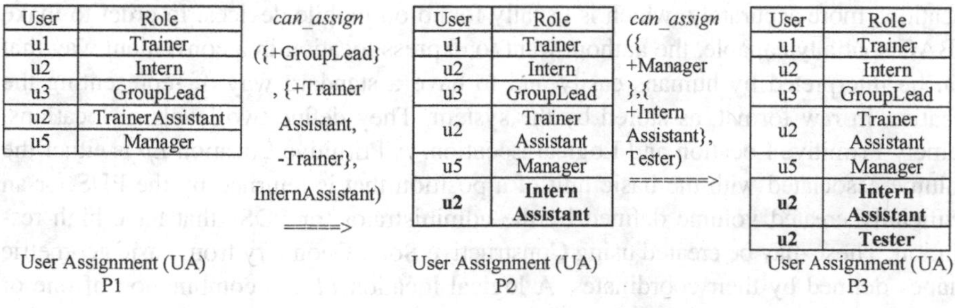

User	Role
u1	Trainer
u2	Intern
u3	GroupLead
u2	TrainerAssistant
u5	Manager

User Assignment (UA)
P1

can_assign
({+GroupLead}
, {+Trainer
Assistant,
-Trainer},
InternAssistant)
=====>

User	Role
u1	Trainer
u2	Intern
u3	GroupLead
u2	Trainer Assistant
u5	Manager
u2	**Intern Assistant**

User Assignment (UA)
P2

can_assign
({
+Manager
},{
+Intern
Assistant},
Tester)
======>

User	Role
u1	Trainer
u2	Intern
u3	GroupLead
u2	Trainer Assistant
u5	Manager
u2	**Intern Assistant**
u2	Tester

User Assignment (UA)
P3

Fig. 3. Executing rule 2 on policies P2

So, the Manager *u5* can continue to assign the Tester role to *u2* users by executing rule 2 on *P2* policies as shown in Fig. 3.

Finally, in the *P3* policies, user *u2* has both *Intern* and role *Tester* role, which violates the security requirements of the system. In conclusion, executing a series of ARBAC rules can make the RBAC policies violate the security requirements of the system, and there should be a tool to assist policies designers to answer the question of whether the implementation of the ARBAC rules on those policies would result in a violation of system security requirements.

2.5   Spatial-Temporal Role-Based Access Control (STRBAC)

In many scenarios, authorization depends on additional contextual information such as the location of the user and the time of the day. In this case, an intern of an organization should only be authorized to access the information system of a company only in the branch he is working and during working hours such as between 7 am and 11 am. In order to understand the authorization conditions that depend on spatial-temporal constraints, we need to introduce the model of location and time.

In [20, 21], the TRBAC model of time is usually specified by means of intervals periodically repeating time intervals, such as day and night-time (two intervals repeating daily), each hour per day (twenty-four intervals repeating daily), or each day per week (seven intervals repeating weekly). Let *TMAX* be a positive integer and *a* is a non-negative integer such that $a + 1 \leq TMAX$, A time slot is a pair *(a; a + 1)*; to ease the readability, we will use (8 am; 4 pm), (4 pm; 12 am), and (12 am; 8 am) to denote time slots (0; 1), (1; 2), and (2; 3), respectively. The set of all time slots is $TSTMAX = \{(a; a + 1) \mid 0 \leq a < TMAX\}$. We will usually write *TS* in place of *TSTMAX* when *TMAX* is clear from context and in this research, we assume *TMAX* to be given so that the set *TS* is fixed. A time instant is a non-negative real number. A time instant *t* belongs to a time slot *(a; a + 1)*, written as $t \in (a; a + 1)$, if and only if $a \leq (t \bmod TMAX) < a + 1$ where *mod* is the usual modulo operator, i.e., $t0 = t \bmod TMAX$ if and only if there exists a non-negative integer *k* such that $t = t0 + k \cdot TMAX$.

The location of a user proposed in [22] should be updated automatically using a position determination system (PDS). GPS is one of the most well-known methods to get locations using satellite. Another method requires infrared sensors base station, infrared transponders and active infrared badges that can respond to the sensors to detect and inform user location to the base station in small organizations. Other methods use wireless signal strength information from multiple stations to estimate the locations more accurately, which is usually found on mobile devices. In order to make RBAC spatially capable, the authors want to express location in a convenient way that can be interpreted by humans easily and to have a standard way of representing the location in raw format, as stored by the system. They define two levels of locations, namely Primitive Location and Logical Location. A Primitive Location *Lp* is either the volume associated with the basic unit of a position that is returned by the PDS, or an artificially created volume defined by the administrator for PDSs that have high resolution. These may be created using Constructive Solid Geometry from basic geometric shapes defined by their coordinates. A logical location *Ll* is a combination of one or more logical or raw locations joined by a ∪, ∩ and / or \ operator combined with other primitive locations to form a logical location. For the sake of simplicity, we will focus on logical location and assume that the location Ll to be updated by the PDS.

An enhanced version of STRBAC is the ESTRBAC model [23], this model proposes new concepts of role extent and permission extent to define the spatial-temporal access control policies. ESTARBAC still consists components of RBAC, namely, users, roles and permissions but they are associated with either spatial extents or spatial-temporal extents. In ESTRBAC model, a set I of intervals is a set of all time slots that participate in at least one spatial-temporal access control policy specification (e.g., I is a subset of TS. For simplicity, we consider $I = TS$ in this research, i.e., every time slot participating in policy specification). Roles and permissions can be available only at specific locations and during specific time intervals, namely, Role Extents (RE) and Permission Extents (PE). In this research, we will use these RE and PE to support our analysis.

In this research, we assume that both L_l and $TMAX$ are given so that the set TS of all time slots is fixed and the set UL of all user logical locations $UL \subseteq U \times L$ is updated from the PDS. STRBAC extends RBAC by adding the Role Extents relation $RE \subseteq R \times L \times TS$, the Permission Extends $PE \subseteq P \times L \times TS$ and replacing the user-role assignment UA with its spatial-temporal user-role assignment relation $UA \subseteq U \times R$ $L \times TS$. For the sake of simplicity, following [24], we exclude role hierarchies.

An extent is a pair (l, ts) which associates spatial-temporal extent to roles or permissions. A role r is enabled at logical location l and time instant t if and only if there exists a time interval ts such that t belongs to ts and $ts \in TS$ and $(r, l, ts) \in RE$. A user u is a member of role r at location l and interval ts if and only if r is enabled at location l and interval ts and $(u; r; l, ts) \in UA$. A user u can activate role r at location l and interval ts if and only if u is a member of role r within extent (l, ts) and u is at location l: $(u, l) \in UL$ and the current time-slot is ts. Similarly, a permission p is enabled at location l and interval ts if and only if $(p; l, ts) \in PE$. A user u has permission p at location l and interval ts if and only if there exists role r such that $(p; r)$ $\in PA$ and p is enabled within extent (l, ts) and u is a member of r within extent (l, ts). A user u can access permission p at location l and interval ts if and only if u has permission p within extent (l, ts) and u is at location l: $(u, l) \in UL$ and the current time-slot is ts. Our STRBAC policies is a tuple $(U; R; P; UA; PA; L; TS; UL; RE; PE)$.

In Fig. 4, the role *Manager* is enabled for users in the *A1 building* at 8:00 am–11:00 am (RE), if Alice's current condition matches the location at *A1 building* at time 8:00 am–11:00 am (UA), Alice can grant the role *Manager*. *Write o1* is activated for the role in *A1* building at 8:00 am–11:00 am (PE). If role *Manager*'s current condition matches the location at *A1 building* at 8:00 am–11:00 am (PA), the role *Manager* will grant *Write o1* permission. Since Alice has the role *Manager* and the role *Manager* has the right to *Write o1* during the time mentioned (8:00 am–11:00 am), Alice has the permission to *Write o1*. Since Bob has the role *Engineer* and the role *Engineer* only has to *Write o2* permission between 1:00 pm–5:00 pm, different from the required times (8:00 am–11:00 am), Bob has no *Write o2* permission.

User	Role	Location	Interval
Alice	Manager	A1 building	Morning(8:00am - 11:00am)
Bob	Engineer	A2 building	Morning (8:00am - 11:00am)

User Assignment (UA)

Role	Permission
Manager	Write o1
Engineer	Write o2
Technician	Read o2

Permission Assignment (PA)

Role	Location	Interval
Manager	A1 building	Morning(8:00am - 11:00am)
Engineer	A1 building	Morning (8:00am - 11:00am)
Engineer	A2 building	Morning (8:00am - 11:00am)

Role Extent (RE)

Permission	Location	Interval
Read o2	A3 building	Afternoon(1:00pm – 5:00pm)
Write o2	A2 building	Afternoon(1:00pm – 5:00pm)
Write o1	A1 building	Morning(8:00am - 11:00am)

Permission Extent (PE)

User	Location	User
Alice	A1 building	Location
Bob	A2 building	(UL)

Fig. 4. STRBAC

2.6 Administrative Spatial-Temporal Role-Based Access Control (ASTRBAC)

One of the Administrative model designed to manage the change of STRBAC policies named ADMINESTAR [9], which allows multiple administrators to modify the STRBAC policies while ensures they cannot abuse the system using their powers. An administrative action consists two components, administrative policies and administrative operation, to define which administrators are allowed to modify ESTRBAC policies. Administrative Policies governs a set of administrative rules to specify which administrative role is authorized to modify ESTARBAC entities of which regular role range. From now on, we focus on the set of administrative rules. All ESTARBAC entities together define the system state, which changes when one or more of the entities change. Administrative Operations are the change of the system state upon their completion only if the administrative policies allow. Administrative Policies and Administrative Operations become more complex if their access control has more attributes.

Our model in [30] based on ADMINESTAR [9], has more constraints in administrative rule. In ADMINESTAR, the administrator condition only has one role so it cannot express actions that require the administrator to have more than one role. In our ASTRBAC model, administrator rule is a set of roles so that administrative actions can describe the more administrative scenario. ASTRBAC focuses on managing data of these entities: *UA, RE, PE, PA* by providing actions on them. These actions are divided into four groups depending on their target, *can_assign_UA*, and *can_revoke_UA* are designed to manage entity *UA*; *can_assign_PA* and *can_revoke_PA* are designed to manage entity *PA*; *can_add_RE* and *can_delete_RE* are designed to manage entity *RE*; and *can_add_PE* and *can_delete_PE* are designed to manage entity *PE*.

We assume that entity *UL* is automatically managed by PDS so there are no actions to manage *UL* in this research; that the basic entities *R, P, L,* and *TS* are finite and constant; and that entity U is infinite. Thus, the STRBAC policies depend on the entities

UA, RE, PE, PA. If one of those entities is modified, the STRBAC state will be changed. Hence, the administrative actions need to be examined carefully as these actions can lead the STRBAC policies to a state in which the security requirement of the system is violated. Such problem is well-known as the reachability problem [10, 11].

In the following, let $\alpha = (U; R; P; UA; PA; L; TS; UL; RE; PE)$ be the STRBAC policies. A signed role is an expression of the form $+r$ or $-r$. A role condition is a finite set of signed roles. A signed role σ in a condition C is positive when there exists a role r such that $\sigma = +r$, a condition C is negative when there exists a role r such that $\sigma = -r$. An administrative action is a tuple $(\{A_{rule}, l_a, ts_a\}, \{R_{rule}, l_u, ts_u\}, Ud)$ where tuple $\{A_{rule}, l_a, ts_a\}$ is called admin pre-condition, A_{rule} is a role condition, (l_a, ts_a) are location and time-slot that together describe spatial temporal constraint on A_{rule}; Tuple $\{R_{rule}, l_u, ts_u\}$ is called user pre-condition where R_{rule} is a role condition; (l_u, ts_u) are location and time-slot that express spatial temporal constraint on R_{rule} that are used together to limit the users whose extents can be modified by the administrator; the Ud element can be an element or many elements depend on each action. The user pre-condition is optional while admin pre-condition is compulsory for all actions.

2.7   ASTRBAC Pre-condition

The admin pre-condition is passed if at least, one administrator satisfies tuple $\{A_{rule}, l_a, ts_a\}$, the positive roles $+r$ in A_{rule} specify the roles administrators must activate at location la during time slot ts_a, the negative roles $-r$ in A_{rule} describe the roles administrators cannot activate within the extent (l_a, ts_a). Administrator a can activate role r within the extent (l, ts) if and only if the formula $\exists tscur: [(ad, l) \in UL \wedge (ad, r, l, ts) \in UA \wedge (r, l, ts) \in RE \wedge (ts_{cur} = ts)]$ returns true, where ts_{cur} is the current time-slot determined by system. The following *check_role_admin* formula ensures the admin pre-condition is checked, if it returns true, there exist an administrator can perform the corresponding action, otherwise, the action is rejected since there is no administrator who can satisfy admin pre-condition.

$check\_role\_admin\ (A_{rule}, l_a, ts_a): \exists\ ad, ts\ [((ad, la) \in UL) \wedge (ts_{cur} = ts_a)$
$\wedge (role \in A_{rule})\ ((ad, r, la, ts_a) \in UA \wedge (r, la, ts_a) \in RE)\ if\ role = +r$
$\wedge (role \in A_{rule})\ ((ad, r, la, ts_a) \notin UA \vee (r, la, ts_a) \notin RE)\ if\ role = -r]$

The user pre-condition $\{R_{rule}, l_u, ts_u\}$ limits which users whose extents can be modified by administrators. The positive roles $+r$ in Rrule specify the roles user must be assigned within extent (l_u, ts_u), negative roles $-r$ specify the roles users must not be assigned within extent (l_u, ts_u). The User Role Assignment relation UA needs to be checked if user u satisfies (R_{rule}, l_u, ts_u) using *check_role_user* formula

$check\_role\_user(u, R_{rule}, l_u, ts_u): \bigwedge\ (role \in R_{rule})\ ((u, r, l_u, ts_u) \in UA\ if\ role = +r)$
$\wedge (role \in R_{rule})\ ((u, r, l_u, ts_u) \notin UA\ if\ role = -r)$

If *check_role_user* returns true, extents (role or permission extents) of user u can be updated, otherwise, it cannot be updated with this action. In conclusion for this pre-condition, an action can be performed if and only if admin pre-condition is passed and there exist a user can be updated (in some actions).

2.8 Administrative Actions

The STRBAC policies have four main sets, set *UA, RE, PE,* and *PA*. The corresponding action for these sets is listed below.

- *Can_Assign_UA (A_{rule}, l_a, ts_a, R_{rule}, l_u, ts_u, r)*
  Can_Revoke_UA (A_{rule}, l_a, ts_a, R_{rule}, l_u, ts_u, r)
- *Can_Add_RE (A_{rule}, l_a, ts_a, r, l, ts)*   *Can_Delete_RE (A_{rule}, l_a, ts_a, r, l, ts)*
- *Can_Assign_PA (A_{rule}, l_a, ts_a, r_t, p_t)*   *Can_Revoke_PA (A_{rule}, l_a, ts_a, r_t, p_t)*
- *Can_Add_PE (A_{rule}, l_a, ts_a, p, l, ts)*   *Can_Delete_PE (A_{rule}, l_a, ts_a, p, l, ts)*

As we only focus on User-Reachability Problem, we will use 4 main administrative actions: *Can_Assign_UA, Can_Revoke_UA, Can_Add_RE, Can_Delete_RE.*

User-Assignment Actions: such actions are designed to manage the actions add or delete tuples in relation UA using role assignment and role revocation actions of users within certain spatial-temporal extents.

Can_Assign_UA (A_{rule}, l_a, ts_a, R_{rule}, l_u, ts_u, r), where *{A_{rule}, l_a, ts_a}* is admin precondition, *[R_{rule}, l_u, ts_u]* is user pre-condition, *(l_u, ts_u, r)* is the update tuple. A_{rule} and R_{rule} can contain positive roles +r in companion within extent *(l_u, ts_u)*, negative roles is −r conflict within extent *(l_u, ts_u)*. Notice that, this action can assign *(l_u, ts_u, r)* to any users that satisfy user pre-condtion. *Can_Assign_UA* is enabled if admin pre-condition is satisfied and there exists a user *u* satisfy user pre-condition R_{rule}.

∃ u (check_role_admin (A_{rule}, l_a, ts_a) ∧ check_role_user(u, R_{rule}, l_u, ts_u))

Once this rule is passed, the tuple *(u, r, l_u, ts_u)* is added to *UA using UA' = UA ∪ (u, r, l_u, ts_u)* where *u* is the user need to assign new roles.

Can_Revoke_UA (A_{rule}, l_a, ts_a, R_{rule}, l_u, ts_u, r) where *(A_{rule}, l_a, ts_a)* is admin precondition, *(l_u, ts_u, r)* is update tuple. R_{rule} can contain positive roles +r in companion within extent *(l_u, ts_u)*, negative roles are −r conflict within extent *(l_u, ts_u)*. *Can_Revoke_UA* revoke *(l_u, ts_u, r)* from any users. The administrator needs to satisfy the admin pre-condition to enable this rule and there exists a user *u* satisfy user precondition R_{rule}, this check uses formula *check_role_admin(A_{rule}, l_a, ts_a)*. Once this rule is passed, the tuple *(u, l_u, ts_u, r)* is removed from *UA: ∃u UA' = UA\(u, r, l_u, ts_u)* where *u* is the user need to revoke user roles.

RoleExtent Actions: such actions are designed to manage the actions add or delete tuples in relation *RE* using adding and deleting actions of role within certain spatial-temporal extents. Role extents that are registered in *RE* can be assigned to users in entity *UA*. In order to activate a role within a spatial-temporal extent, that role must be assigned to the user in *UA* and enable within that extent in *RE*.

Can_Add_RE (A_{rule}, l_a, ts_a, r, l, ts) where *(A_{rule}, l_a, ts_a)* is admin pre-condition, *(r, l, ts)* is the update tuple. The admin pre-condition must be checked using the formula: *check_role_admin (A_{rule}, l_a, ts_a)*. Once this rule is passed, the spatial-temporal extent of role *r* is added with tuple *(l, ts): RE' = RE ∪ (r, l, ts)*.

Can_Delete_RE (A_{rule}, l_a, ts_a, r, l, ts) where *(A_{rule}, l_a, ts_a)* is admin pre-condition, *(r, l, ts)* is the update tuple. The admin pre-condition must be checked to enable this action using the formula: *check_role_admin (A_{rule}, l_a, ts_a)*. Once this rule is passed, the spatial-temporal extent of role *r* is deleted with tuple *(l, ts): RE' = RE\(r, l, ts)*.

User	Role	Location	Interval
Alice	manager	A1 building	Morning(8:00am - 11:00am)
Bob	engineer	A2 building	Morning (8:00am - 11:00am)
Peter	technician	A3 building	Afternoon(1:00pm – 5:00pm)
Shan	Engineer	A1 building	Afternoon(1:00pm – 5:00pm)

User Assignment (UA)

Role	Permission
manager	Write o1
engineer	Write o2
technician	Read o2
manager	Write o2

Permission Assignment (PA)

Role	Location	Interval
Manager	A2 building	Morning (8:00am - 11:00am)
Engineer	A1 building	Morning (8:00am - 11:00am)
Engineer	A2 building	Morning (8:00am - 11:00am)
Manager	A1 building	Morning (8:00am - 11:00am)
Technician	A3 building	Afternoon (1:00pm – 5:00pm)

Role Extent (RE)

Permission	Location	Interval
Read o2	A3 building	Afternoon (1:00pm – 5:00pm)
Write o2	A2 building	Morning (8:00am - 11:00am)
Write o1	A1 building	Afternoon (1:00pm – 5:00pm)

Permission Extent (PE)

User	Location	User Location (UL)
Alice	A1 building	
Bob	A2 building	
Shan	A1 building	
Peter	A3 building	

Fig. 5. STRBAC – P1

A run of an ASTRBAC system $(\alpha_0; \varphi)$ is a (possibly infinite) sequence $(\alpha_0; 0) \dots (\alpha_i; t_i); (\alpha_{i+1}; ti+1), \dots$ of states such that $(\alpha_i; t_i) \rightarrow (\alpha_{i+1}; t_{i+1})$ and $t_i \leq t_{i+1}$ for $i = 1 \dots n - 1$ with $n > 1$. If the run is finite, i.e. it is of the form $(\alpha_0; 0) \dots (\alpha_n; t_n)$ for some $n \geq 0$, we say that $(\alpha_n; t_n)$ is the final state of the run.

2.9 User-Role Reachability for STRBAC

Even if administrators can only execute a given set of administrative actions mentioned above, it is still very difficult to foresee all possible interleaving of actions when many administrators perform their administrative actions together with their effect on the initial STRBAC policies. Therefore, in some cases, an untrusted user can gain, in some spatial-temporal, a permission which that person should not gain. In order to identify this situation, we need to solve the next analysis problem.

A reachability problem for an ASTRBAC system $(\alpha_0; \varphi)$ is identified by a tuple $(u; C_f; l_f, ts_f)$ and a set of administrator actions φ to check if there exists a finite run of the ASTRBAC system whose final state $(\alpha_f; l_f, ts_f)$ is such that user u, location l_s and timeslot ts_f satisfy condition C_f under UA_f and l_s, ts_f satisfies C_f under RE_f, and l_s, ts_f satisfies C_f under PE_f where $\alpha_f = (UA_f; RE_f; PE_f)$.

Example 1: Similar to the ARBAC model, here is an example of a system that has policies *P1* as shown in Fig. 5, which has the security requirement (write o1, A1, afternoon), and a set of ARBAC rules as follows:

1. *Can_Update_ UL (Marry, A2)*
2. *Can_Assign_PA ({manager}, A1, morning, engineer, write o1)*

3. *Can_Assign_UA({manager}, A1, morning, {engineer, −technician, −manager}, tester, A2, afternoon)*

4. *Can_Add_RE ({manager}, A1, morning, engineer, A1, afternoon)*

5. *Can_Add_RE({manager, A1, morning, tester, A2, afternoon)*

6. *Can_Add_PA({engineer}, A1, morning, tester, read o2)*

After that, Alice can execute Administrative Rule 2 *Can_Assign_PA ({manager}, A1, morning, engineer, write o1)* to add the line *(engineer, Write o1)* to the *PA* table. After implementing this rule, policies *P1* will be transformed as shown in Fig. 6.

Next, Alice can execute Rule 4 *Can_Add_RE ({manager}, A1, morning, engineer, A1, afternoon)*. After applying this rule, policies *P1* will be changed as in Fig. 7. Finally, using the policies *P3*, user *Shan* who has the role *Engineer* can *write o1*, which violates the original security requirement of the system.

User	Role	Location	Interval
Alice	*manager*	*A1 building*	*Morning (8:00am - 11:00am)*
Bob	engineer	A2 building	Morning (8:00am - 11:00am)
Peter	techni-cian	A3 building	Afternoon (1:00pm − 5:00pm)
Shan	engineer	A1 building	Afternoon (1:00pm − 5:00pm)

User Assignment (UA)

Role	Permission
manager	Write o1
engineer	Write o2
technician	Read o2
manager	Write o2
engineer	**Write o1**

Permission Assignment (PA)

Role	Location	Interval
Manager	A2 building	Morning (8:00am - 11:00am)
Engineer	A1 building	Morning (8:00am - 11:00am)
Engineer	A2 building	Morning (8:00am - 11:00am)
Manager	*A1 building*	*Morning (8:00am - 11:00am)*
Technician	A3 building	Afternoon (1:00pm − 5:00pm)

Role Extent (RE)

Permis-sion	Loca-tion	Interval
Read o2	A3 build-ing	Afternoon (1:00pm − 5:00pm)
Write o2	A2 build-ing	Morning (8:00am - 11:00am)
Write o1	A1 build-ing	Afternoon (1:00pm − 5:00pm)

Permission Extent (PE)

User	Location	User
Alice	*A1 building*	Location
Bob	A2 building	(UL)
Shan	A1 building	
Peter	A3 building	

Fig. 6. STRBAC – P2

3 Automated Analysis for ASTRBAC

At first, we translate ASTRBAC policies to First-Order Logic formula which belongs to Bernays-Schonfinkel-Ramsey (BSR) [27] class to determine the satisfy-ability of formula, which has the form $\exists \underline{x}. \forall \underline{y}.\varphi (\underline{x}; \underline{y})$, where φ is a quantifier-free formula,

x and y are (disjoint and possibly empty) tuples of variable. After that, we use Model Checking Modulo Theories (MCMT) [25], which is a framework to solve reachability problem for infinite state systems that can be represented by transition systems whose set of states and transitions are encoded as constraints in First-Order Logic. MCMT framework uses a backward reachability procedure to solve a particular class of constraint satisfy-ability problems, called Satisfy-ability Modulo Theories (SMT) problems. According to [26], MCMT framework is a scalable and efficient SMT solver currently available.

User	Role	Location	Interval
Alice	manager	A1 building	Morning (8:00am - 11:00am)
Bob	engineer	A2 building	Morning (8:00am - 11:00am)
Peter	technician	A3 building	Afternoon(1:00pm – 5:00pm)
Shan	engineer	A1 building	Afternoon(1:00pm – 5:00pm)

User Assignment (UA)

Role	Permission
Manager	Write o1
Engineer	Write o2
technician	Read o2
Manager	Write o2
Engineer	**Write o1**

Permission Assignment (PA)

Role	Location	Interval
Manager	A2 building	Morning (8:00am - 11:00am)
Engineer	A1 building	Morning (8:00am - 11:00am)
Engineer	A2 building	Morning (8:00am - 11:00am)
Manager	A1 building	Morning (8:00am - 11:00am)
Technician	A3 building	Afternoon (1:00pm – 5:00pm)
Engineer	**A1 building**	**Afternoon (1:00pm – 5:00pm)**

Role Extent (RE)

Permission	Location	Interval
Read o2	A3 building	Afternoon(1:00pm – 5:00pm)
Write o2	A2 building	Morning (8:00am - 11:00am)
Write o1	A1 building	Afternoon(1:00pm – 5:00pm)

Permission Extent (PE)

User	Location	User Location (UL)
Alice	A1 building	
Bob	A2 building	
Shan	A1 building	
Peter	A3 building	

Fig. 7. STRBAC – P3

3.1 Implementing the Translator

We implement our technique which will be discussed below in a tool called ASASPSPACETIME (Automated Symbolic Analysis of Security Policies tool with SPACE and TIME). As in Fig. 8, this tool has two main parts, the Translator, implemented in Python, will get the input of our ASTRBAC reachability problem *(u; C; l; ts)* as reachability problem for STRBAC policies *(α_0; φ)*, and translate all the ASTRBAC policy and Goal to BSR-STS. The second part of the analysis of ASTRBAC policies uses SMT-based model checker named MCMT [25] to solve our problem. Based on the result from MCMT, our tool will answer our problem with statement "reachable" or "unreachable" and show the sequence of actions which changed our STRBAC policies from α_o to α_n where α_n can satisfy *(u; C; l; ts)*. According to [27, 28], we try to reduce our reachability problem for ASTRBAC model to a (finite) sequence of constraint satisfaction problem.

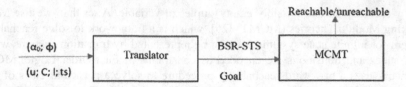

Fig. 8. Our technique to solve the reachability problem for ASTRBAC

Here is how we translate the ASTRBAC to First-Order Logic in BSR class. We need to translate an initial state, the administrative actions, time passing, and the goal.

- **Our state variable** in ASTRBAC contains *re, ua, loc,* and *at* where *re* is a variable of for the current role extent *RE*, similarly, *ua* for *UA, loc* for *UL* and *at* is current system time.
- **Our initial state** contains the tuple $\alpha_0 = (RE, UA, UL, ts_0)$ where *ts* is timeslot, which can be translated as below.

$$
\begin{aligned}
&\forall x,y,z,t.\ ua(x,y,z,t) \iff V_{(u,r,l,ts)\ \in UA}\,(x{=}u \wedge y{=}r \wedge z{=}l \wedge t{=}ts) \wedge \\
&\qquad re(y, z, t) \iff V_{(r,l,ts)\ \in RE}\,(y{=}r,\ z{=}l,\ t{=}ts) \wedge \\
&\qquad at(z) \iff z{=}ts_o
\end{aligned}
$$

Example 2: We analyze a simple example of ASTRBAC

Let U = {Alice; Bob; Peter}
R = {Manager; Engineer; Technician; Tester; Developer}
L = {A1 building; A2 building}
I = {Morning: [8:00 am–12:00 pm]; Afternoon: (12:00 pm–18:00 pm), Night: (18:00 pm–8:00 am)}
UA = {(Alice, Manager, A1 building, Morning); (Bob, Engineer, A2 building, Morning);}
RE = {(Engineer, A1 building, Morning: [8:00 am–12:00 pm]), (Technician, A2 building, Afternoon: [12:00 pm–18:00 pm])}
UL = {(Alice, A1 building); (Bob, A2 building)}
Current time = 8am;

The ASTRBAC rule contains these rule.

1. *Can_Assign_UA({manager}, a1 building, morning, {engineer}, a2 building, afternoon, tester)*
2. *Can_Revoke_UA({manager}, a1 building, morning, {engineer}, a2 building, afternoon, tester)*
3. *Can_Add_RE ({manager}, a1 building, morning, engineer, a1 building, afternoon)*
4. *Can_Delete_RE ({manager}, a1 building, morning, engineer, a1 building, afternoon)*

According the example above, the current time belongs to time slots *Morning*, so our initial state will be

$\forall x,y,z,t.\ ua(x,y,z,t) \Leftrightarrow$
((x= Alice ∧ y= Manager ∧ z= A1 building ∧ t= Morning) ∨ (x= Bob ∧ y= Engineer ∧ z= A2 building ∧ t= Afternoon)) ∧
  re(y, z, t) ⇔ ((y = Engineer, z= A1 building, t= Morning) ∨ (y = Technician, z= A2 building, t= Afternoon)) ∧
  at(z) ⇔ z=Morning

Our ASTRBAC now contains 4 actions *(Can_Assign_UA, Can_Revoke_UA, Can_Add_RE, Can_Delete_RE)* and a goal. We translate each of them as follow.

- **Can_Assign_UA $(A, l_a, ts_a, R_u, l_u, ts_u, r_u)$**

$\Leftrightarrow \exists u_a, u, ts.\ at(ts) \wedge ts = ts_a \wedge$
  $Loc(u_a, l_a) \wedge$
  $\wedge_{r \in A} re(r, l_a, t_a) \wedge \wedge_{r \in A} \neg re(r, l_a, t_a) \wedge$
  $\wedge_{r \in A} ua(u_a, r, l_a, t_a) \wedge \wedge_{r \in A} \neg ua(u_a, r, l_a, t_a) \wedge$
  $\wedge_{r \in R} ua(u, r, l_u, t_u) \wedge \wedge_{r \in R} \neg ua(u, r, l_u, t_u) \wedge$
  $(\forall_{x,y,z,t} ua'(x, y, z, t) \Leftrightarrow ua(x, y, z, t) \vee (x = u \wedge y = r_u \wedge z = l_u \wedge t = ts_u)) \wedge$
  $(\forall_{x,y,z,t} re'(y, z, t) \Leftrightarrow re(y, z, t)) \wedge$
  $(\forall_{x,y,z,t} loc'(x, z) \Leftrightarrow loc(x, z))$

Example 3: According to Example 2, the *Can_Assign_UA({Manager}, A1 building, Morning, {Engineer}, A2 building, Afternoon, Tester)* can be translated as

$\exists u_a, u, ts.\ at(ts)$
at(ts) ∧ ts = Morning ∧
  Loc(Manager, A1 building) ∧
  re(Manager, A1 building, Morning) ∧
  ua(u_a, Manager, A1 building, Morning) ∧
  ua(u, Engineer, A2 building, Afternoon) ∧
  $(\forall_{x,y,z,t} ua'(x, y, z, t) \Leftrightarrow ua(x, y, z, t)$
∨ (x = u ∧ y = Tester ∧ z = A2 building ∧ t = Afternoon)) ∧
  $(\forall_{x,y,z,t} re'(y, z, t) \Leftrightarrow re(y, z, t)) \wedge$
  $(\forall_{x,y,z,t} loc'(x, z) \Leftrightarrow loc(x, z))$

- **Can_Revoke_UA $(A, l_a, ts_a, R, l_u, ts_u, r_u)$**
$\Leftrightarrow \exists u_a, u, ts.\ at(ts) \wedge ts = ts_a \wedge$
  $Loc(u_a, l_a) \wedge$
  $\wedge_{r \in A} re(r, l_a, t_a) \wedge \wedge_{r \in A} \neg re(r, l_a, t_a) \wedge$
  $\wedge_{r \in A} ua(u_a, r, l_a, t_a) \wedge \wedge_{r \in A} \neg ua(u_a, r, l_a, t_a) \wedge$
$\wedge_{r \in R} ua(u, r, l_u, t_u) \wedge \wedge_{r \in R} \neg ua(u, r, l_u, t_u) \wedge$
  $(\forall_{x,y,z,t} ua'(x, y, z, t) \Leftrightarrow ua(x, y, z, t) \wedge \neg (x = u \wedge y = r_u \wedge z = l_u \wedge t = ts_u))$
$\wedge$
  $(\forall_{x,y,z,t} re'(y, z, t) \Leftrightarrow re(y, z, t)) \wedge$
  $(\forall_{x,y,z,t} loc'(x, z) \Leftrightarrow loc(x, z))$

Example 4: According to Example 2, the *Can_Revoke_UA({Manager}, A1 building, Morning, {Engineer}, A2 building, Afternoon, Tester)* can be translated as

$\exists u_a,\ u,\ ts.\ at(ts)$
$at(ts) \land ts = Morning \land$
  $Loc(Manager, A1\ building) \land$
  $re(Manager, A1\ building, Morning) \land$
  $ua(u_a, Manager, A1\ building, Morning) \land$
  $ua(u, Engineer, A2\ building, Afternoon) \land$
  $(\forall_{x,\ y,\ z,\ t}\ ua'(x, y, z, t) \Leftrightarrow ua(x, y, z, t)$
$\land \neg\ (x = u \land y = Tester \land z = A2\ building \land t = Afternoon)) \land$
  $(\forall_{x,\ y,\ z,\ t}\ re'(y, z, t) \Leftrightarrow re(y, z, t)) \land$
  $(\forall_{x,\ y,\ z,\ t}\ loc'(x, z) \Leftrightarrow loc(x, z))$

- **Can_Add_RE $(A, l_a, ts_a, r_u, l_u, ts_u)$**
$\Leftrightarrow \exists u_a,\ u,\ ts.\ at(ts) \land ts = ts_a \land$
  $Loc(u_a, l_a) \land$
  $\land_{+r \in A}\ re(r, l_a, t_a) \land \land_{r \in A} \neg re(r, l_a, t_a) \land$
  $\land_{+r \in A}\ ua(u_a, r, l_a, t_a) \land \land_{r \in A} \neg ua(u_a, r, l_a, t_a) \land$
  $(\forall_{x,\ y,\ z,\ t}\ ua'(x, y, z, t) \Leftrightarrow ua(x, y, z, t) \land$
  $(\forall_{x,\ y,\ z,\ t}\ re'(y, z, t) \Leftrightarrow re(y, z, t)) \lor (y = r_u \land z = l_u \land t = ts_u)) \land$
  $(\forall_{x,\ y,\ z,\ t}\ loc'(x, z) \Leftrightarrow loc(x, z))$

Example 5: According to Example 2, the *Can_Add_RE ({Manager}, A1 building, Morning, Engineer, A1 building, Afternoon)* can be translated as

$\exists u_a,\ u,\ ts.\ at(ts) \land ts = Manager \land$
  $Loc(A1\ building, Morning) \land$
  $re(Manager, A1\ building, Morning) \land$
  $ua(u_a, Manager, A1\ building, Morning) \land$
  $(\forall_{x,\ y,\ z,\ t}\ ua'(x, y, z, t) \Leftrightarrow ua(x, y, z, t) \land$
  $(\forall_{x,\ y,\ z,\ t}\ re'(y, z, t) \Leftrightarrow re(y, z, t))$
$\lor (y = Engineer \land z = A1\ building \land t = Afternoon)) \land$
  $(\forall_{x,\ y,\ z,\ t}\ loc'(x, z) \Leftrightarrow loc(x, z))$

- **Can_Delete_RE $(A, l_a, ts_a, r_u, l_u, ts_u)$**
$\Leftrightarrow \exists u_a,\ u,\ ts.\ at(ts) \land ts = ts_a \land$
  $Loc(u_a, l_a) \land$
  $\land_{+r \in A}\ re(r, l_a, t_a) \land \land_{r \in A} \neg re(r, la, ta) \land$
  $\land_{+r \in A}\ ua(ua, r, la, ta) \land \land_{r \in A} \neg ua(ua, r, la, ta) \land$
  $(\forall_{x, y, z, t}\ ua'(x, y, z, t) \Leftrightarrow ua(x, y, z, t) \land$
  $(\forall_{x, y, z, t}\ re'(y, z, t) \Leftrightarrow re(y, z, t)) \land \neg (y = ru \land z = lu \land t = tsu)) \land$
  $(\forall_{x, y, z, t}\ loc'(x, z) \Leftrightarrow loc(x, z))$

Example 6: According to Example 2, the *Can_Delete_RE ({Manager}, A1 building, Morning, Engineer, A1 building, Afternoon)* can be translated as

$\exists u_a, u, ts. at(ts) \land ts = Manager \land$
    Loc(A1 building, Morning) $\land$
    re(Manager, A1 building, Morning) $\land$
    $ua(u_a,$ *Manager, A1 building, Morning)* $\land$
    $(\forall_{x, y, z, t}$ *ua'(x, y, z, t)* $\Leftrightarrow$ *ua(x, y, z, t)* $\land$
    $(\forall_{x, y, z, t}$ *re'(y, z, t)* $\Leftrightarrow$ *re(y, z, t))* $\land \neg$ *(y = Engineer*
$\land z = A1$ *building* $\land t = Afternoon))$ $\land$
    $(\forall_{x, y, z, t}$ *loc'(x, z)* $\Leftrightarrow$ *loc(x, z))*

Time passing: as mentioned before, the following formula means that every time the state of the time change, we will move to the next time slots.

If $j + 1 < T_{max}$ then

$at(ts) \land ts = (j, j+1) \land$
$\forall y, z, t\ re' \Leftrightarrow re(y, z, t) \land \forall x, y, z, t.\ ua'(x, y, z, t) \Leftrightarrow ua\ (x, y, z, t) \land$
$\forall z.\ L'(z) \Leftrightarrow L(z) \land \forall t.\ at'(t) \Leftrightarrow t = (j+1, j+2)$

Otherwise:

$at(ts) \land ts = (T_{max}-1, T_{max}) \land$
$\forall y, z, t.\ re' \Leftrightarrow re(y, z, t) \land \forall x, y, z, t.\ ua'(x, y, z, t) \Leftrightarrow ua\ (x, y, z, t) \land$
$\forall z.\ L'(z) \Leftrightarrow L(z) \land \forall t.\ at'(t) \Leftrightarrow t = (j+1, j+2)$

Goal state:

$(u_g, r_g, l_g, ts_g) \Leftrightarrow$
$\exists u, r, l, ts.\ re(r, l, ts) \land ua(u, r, l, ts) \land u = u_g \land r = r_g \land l = l_g \land ts = ts_g$

Example 7: Our goal *(Alice, Write_O1, A1 building, Afternoon)* can be translated as

$\exists u, r, l, ts.\ re(r, l, ts) \land ua(u, r, l, ts) \land u = Alice$
$\land r = Write\_O1 \land l = A1$ *building*
$\land ts = Afternoon$

After translating all the ASTRBAC policies to BSR, we need to do an *AND* operator on all the translated formulas.

- If the result returns true, then our goal is reachable, otherwise, it is unreachable.
- If the state is reachable, we know that our system is unsafe, otherwise, it is safe.

Heuristics

As mentioned before, safety analysis of the ASTRBAC system may lead to the state space explosion, which is often seen with model checking approaches. Therefore, this research also focuses on the design and implement heuristics to reduce the state space explosion problem and improve the performance of our solutions. Here is an example of the initial settings and the corresponding administrative actions as follows.

Example 8: We use integers representing roles with the set $R = \{1, 2, ..., 16\}$, set interval $I = \{ts1, ts2, ts3, ts4, ts5\}$, location $L = \{l1, l2, l3, l4, l5\}$ and goal $(1, l2, ts1)$. The ASTRBAC rule contains these rules.

1. *Can_Assign_UA(6, l5, ts4, {true}, l1, ts4, 4)*
2. *Can_Assign_UA(6, l2, ts1, {true}, l3, ts2, 4)*
3. *Can_Assign_UA(true, l4, ts3, {true}, l3, ts2, 4)*
4. *Can_Assign_UA(6, l4, ts2, {9}, l2, ts1, 1)*
5. *Can_Assign_UA(true, l2, ts2, {true}, l4, ts2, 6)*
6. *Can_Assign_UA(6, l4, ts2, {true}, l2, ts1, 9)*
7. *Can_Assign_UA(true, l2, ts1, {true}, l3, ts2, 9)*
8. *Can_Add_RE(true, l2, ts1, {−1 −10}, l4, ts2, 6)*
9. *Can_Add_RE(10, l2, ts2, {−1 −10}, l2, ts2, 9)*
10. *Can_Add_RE(12, l2, ts2, {10}, l2, ts2, 11)*

3.1.1   Forwarding

Forwarding is used to reduce administrative actions need to be checked. From the initial administrator actions, the heuristic starts filtering the administrative actions needed to be check and removing unrelated administrative actions. The initial state in the above example is when no user has been assigned the role or we can say that the *UA* set is empty. Initially, the heuristic starts filtering the administrative actions that can be executed from the initial state in one step by checking if both administrative conditions A and user conditions R_u is satisfied:

- The administrative conditions A will be satisfied if the roles in A are enabled and assigned to the location l_a and timeslot ts_a.
- The user conditions R_u are satisfied if the roles in R are enabled and assigned to the location l_u and timeslots ts_u.

The administrative actions taken in this step are called useful actions in one-step from the initial state. After that, the heuristic calculates states that can be reached after executing useful actions in one step from the initial state. These states are used to filter out useful actions in the next step (useful action in 2 steps from the initial stage). This process is repeated until there is no more administrator to add (the useful actions set is no longer expandable).

Finally, the administrative actions filtered by forwarding can have fewer actions than the original actions since it removed the administrative actions that are not necessary for the checking.

Example 9: We will run the reachability problem in Example 8 by using forwarding. In the first step, the administrative actions were taken:

3. *Can_Assign_UA(true, l4, ts3, {true}, l3, ts2, 4)*
5. *Can_Assign_UA(true, l2, ts2, {true}, l4, ts2, 6)*
7. *Can_Assign_UA(true, l2, ts1, {true}, l3, ts2, 9)*
8. *Can_Add_RE(true, l2, ts1, {−1 −10}, l4, ts2, 6)*

In the first step, action *(3)* is taken because:

- The administrative condition *A* is satisfied: the condition of the action *(3)* is true, ensuring *A* is always satisfied
- The user condition R_u is satisfied: the condition is true, ensuring that R_u is always satisfied.

Similarly for actions *(5)*, *(7)*. Actions *(8)* is taken for:

- Administrators conditions *A* is satisfied: the condition of the action *(8)* is true, ensuring *A* is always satisfied
- The user condition R_u is satisfied: is that the condition *C* of the action *(5.8)* is the negative roles, ensuring that R_u is always satisfied

In the second step, the heuristic will calculate the new states that are likely to be reached:

(4, l3, ts2) after execution *(3)*; *(6, l4, ts2)* after execution *(5)*;
(9, l3, ts2) after execution *(7)*;

and role 6 is enabled at *l4* and *ts2* after execution *(8)*.

Then, the heuristic will take useful actions in the second step by finding possible actions taken from these new states.

3. *Can_Assign_UA(true, l4, ts3, {true}, l3, ts2, 4)*
5. *Can_Assign_UA(true, l2, ts2, {true}, l4, ts2, 6)*
6. *Can_Assign_UA(6, l4, ts2, {true}, l2, ts1, 9)*
7. *Can_Assign_UA(true, l2, ts1, {true}, l3, ts2, 9)*
8. *Can_Add_RE(true, l2, ts1, {−1 −10}, l4, ts2, 6)*

In this step, action *(6)* is taken because:

- The administrative condition *A* is satisfied: the condition of the action *(6)* is satisfied since *role 6* is enabled at *l4* and *ts2* by using administrative action *(8)* and *role 6* is assigned to a user at *l4* and *ts2* by using administrative action *(5)* when performing useful actions in the first step as described above;
- The user condition R_u is satisfied: the condition is true, ensuring that R_u is always satisfied.

In the third step, the heuristic will calculate the new states that are likely to be reached:

(4, l3, ts2) after execution *(3)*; *(6, l4, ts2)* after execution *(5)*;
(9, l2, ts1) after execution *(6)*; *(9, l3, ts2)* after execution *(7)*;

and *role 6* is enabled at *l4* and *ts2* after execution *(8)*.

Then, heuristic will take useful actions in third step by finding possible actions taken from these new states.

3. *Can_Assign_UA(true, l4, ts3, {true}, l3, ts2, 4)*
4. *Can_Assign_UA(6, l4, ts2, {9}, l2, ts1, 1)*
5. *Can_Assign_UA(true, l2, ts2, {true}, l4, ts2, 6)*

6. *Can_Assign_UA(6, l4, ts2, {true}, l2, ts1, 9)*
7. *Can_Assign_UA(true, l2, ts1, {true}, l3, ts2, 9)*
8. *Can_Add_RE(true, l2, ts1, {−1 −10}, l4, ts2, 6)*

In this step, action *(4)* is taken because:

- The administrative condition A is satisfied: the condition of the action *(4)* is satisfied since *role 6* is enabled at *l4* and *ts2* by using administrative action *(8)* as mentioned before, and *role 6* is assigned to a user at *l4* and *ts2* by using administrative action *(5)* when performing useful actions in first step as mentioned before;
- The user condition R_u is satisfied: role *9* is assigned at *l2* and *ts1* by using action *(6)* in second step.

In the fourth step, the heuristic will calculate the new states that are likely to be reached and the possible actions will be *(3), (4), (5), (6), (7), (8)*. However, when heuristic tries to find more useful actions, the number of actions taken are similar to the third step *(3), (4), (5), (6), (7), (8)*. Obviously, the actions cannot be extended anymore (reaching Fix-Point). The administrative actions taken will be used instead of the original actions in Example 8.

3.1.2 Backwarding

This heuristic is similar to forwarding introduced above. However, in this heuristic, the checking process will start from the goal. From the administrative actions that executing it will reach the goal in one step, the heuristic starts filtering administrative actions related to the checking and removing unrelated administrative actions.

At first, the heuristic filters the administrative actions that executing it can reach the goal in one step; the filtered actions should satisfy the same administrative conditions A and user conditions R_u as same as the Forwarding heuristic. Then, backwarding will continue filtering administrative actions that accomplish it will reach the goal in two steps, three steps, ... to n steps until no further administrative actions can be added.

Finally, the filtered administrative actions can be fewer than the original actions since it removed the administrative actions that are not necessary for the checking.

Example 10: We will run the reachability problem in Example 8 by using backwarding. In the first step, the administrative actions were taken:

4. *Can_Assign_UA(6, l4, ts2, {9}, l2, ts1, 1)*

In this step, action *(4)* is taken because:

- The target role *(1, l2, ts2)* which is similar to final goal, thus, executing this action will reach the final goal in one step.

In the second step, the administrative actions were taken:

4. *Can_Assign_UA(6, l4, ts2, {9}, l2, ts1, 1)*
5. *Can_Assign_UA(true, l2, ts2, {true}, l4, ts2, 6)*
6. *Can_Assign_UA(6, l4, ts2, {true}, l2, ts1, 9)*
8. *Can_Add_RE(true, l2, ts1, {−1 −10}, l4, ts2, 6)*

In this step, we see that in order to execute action *(4)*:

- The administrative condition A must be satisfied: the condition of the action *(4)* is satisfied when *role 6* is enabled at *l4* and *ts2* and *role 6* is assigned at *l4* and *ts2*.
- The user condition R_u must be satisfied: the condition of the action *(4)* is satisfied when *role 9* is assigned at *l2* and *ts1*.

In this step, action *(8)* is taken because:

- The condition satisfied the administrative condition of the action *(4)* when *role 6* is enabled at *l4* and *ts2*.
- The administrative condition *A* is satisfied when it is true.
- The user condition R_u is satisfied when it contains negative role.

In this step, action *(5)* is taken because:

- The condition satisfied the administrative condition of the action *(4)* when *role 6* is enabled at *l4* and *ts2*.
- The administrative condition *A* is satisfied when it is true.
- The user condition R_u is satisfied when it is true.

In this step, action *(6)* is taken because:

- The condition satisfied the user condition of the action *(4)* when *role 9* is assigned at *l2* and *ts1*.
- The administrative condition *A* is satisfied when *role 6* is enabled at *l4* and *ts2*.
- The user condition R_u is satisfied when it is true.

In the third step, the heuristic will calculate the new states that are likely to be reached. However, no administrative actions are added. Obviously, the action sets cannot be extended anymore (reaching Fix-Point). The administrative actions taken will be used instead of the original administrative action set in Example 8.

3.1.3   Combining Forwarding and Backwarding
This heuristic filters the intersection of administrative actions obtained after the execution of the Forwarding and Backwarding since these actions are effective in the checking progress.

It is obvious that if the goal is reachable, the set of administrative actions required for the checking from the initial state to the goal state must be within the filtered actions from the Forwarding heuristic and Backwarding heuristic. Filtering the intersection of administrative actions will remove actions that do not contribute to the checking whether the goal has been reached.

In best case scenario, this heuristic can be very useful and can contain less actions from the original actions. In some cases, the result may be empty when the two sets do not intersect, which can return result faster and optimize the runtime.

Example 11: We will run the reachability problem in Example 8 by using the result from forwarding (Example 9) and backwarding (Example 10), the action set obtained after running the heuristic:

Forwarding
3. *Can_Assign_UA(true, l4, ts3, {true}, l3, ts2, 4)*
4. *Can_Assign_UA(6, l4, ts2, {9}, l2, ts1, 1)*
5. *Can_Assign_UA(true, l2, ts2, {true}, l4, ts2, 6)*
6. *Can_Assign_UA(6, l4, ts2, {true}, l2, ts1, 9)*
7. *Can_Assign_UA(true, l2, ts1, {true}, l3, ts2, 9)*
8. *Can_Add_RE(true, l2, ts1, {−1 −10}, l4, ts2, 6)*

Backwarding
4. *Can_Assign_UA(6, l4, ts2, {9}, l2, ts1, 1)*
5. *Can_Assign_UA(true, l2, ts2, {true}, l4, ts2, 6)*
6. *Can_Assign_UA(6, l4, ts2, {true}, l2, ts1, 9)*
8. *Can_Add_RE(true, l2, ts1, {−1 −10}, l4, ts2, 6)*

The intersection of these two sets will be:

4. *Can_Assign_UA(6, l4, ts2, {9}, l2, ts1, 1)*
5. *Can_Assign_UA(true, l2, ts2, {true}, l4, ts2, 6)*
6. *Can_Assign_UA(6, l4, ts2, {true}, l2, ts1, 9)*
8. *Can_Add_RE(true, l2, ts1, {−1 −10}, l4, ts2, 6)*

The administrative actions taken will be used instead of the original administrative action set in the example 8 and even Examples 9 and 10.

3.1.4   Combining Complement Actions

From the administrative action set obtained by the previous heuristic, the heuristic will combine administrator actions pairs to get the equivalent actions and reduce the number of actions,

The heuristic will perform the search for pairs of administrative actions which can combine into equivalent actions. If there are two actions which are combinable, the heuristics will replace both actions, which are not used for the checking, with a new equivalent action.

Example 12: We will run the reachability problem in an example below by using this heuristic to combine complement actions. In this step, the heuristic will search for any complement actions. If there exist 2 actions such as:

11. *Can_Assign_UA(true, l2, ts2, {7, 12}, l2, ts4, 8)*
12. *Can_Assign_UA(true, l2, ts2, {7, −12}, l2, ts4, 8).*

Since these 2 actions are complement, the next step is to combine these 2 actions into *Can_Assign_UA(true, l2, ts2, {7}, l2, ts4, 8)*. The new action will replace both the original actions.

3.1.5 Removing Duality Actions

From the administrative actions obtained by the previous heuristic, this heuristic removes the unneeded actions from the duality administrator actions to reduce the number of actions needs checking.

The first heuristic will perform the search for pairs of duality administrative actions. If there are two duality administrative actions, the heuristics will remove the actions which are not used for the checking.

Example 13: We will run the reachability problem in an example by using this heuristic to remove duality actions. In this step, the heuristic will search for any duality actions.

If there are 2 actions such as:

13. $Can\_Assign\_UA(true, l3, ts1, \{-5, 7\}, l2, ts1, 13)$
14. $Can\_Assign\_UA(true, l3, ts1, \{5\}, l2, ts1, 7)$

In action *(13)*, the user condition R_u require role *{−5}* and role *{7}* to be able to assign role *{13}*. However, in action *(14)*, the user condition R_u require role *{5}* to be able to assign role *{7}*.

If there's no other action to assign role *(7)* and to revoke role *(5)* than we can remove the action *(13)* since action *(14)* is always executable while action *(13)* will never be executed.

If there are 3 actions such as:

15. $Can\_Assign\_UA(true, l3, ts1, \{5, 7\}, l2, ts1, 15)$
16. $Can\_Assign\_UA(true, l3, ts1, \{1 -2\}, l2, ts1, 5)$
17. $Can\_Assign\_UA(true, l3, ts1, \{-1 2\}, l2, ts1, 7)$

In action *(15)*, the user condition R_u require role *{5}* and role *{7}* to be able to assign role *{15}*.

In action *(16)*, the user condition R_u require role *{1}* and role *{−2}* to be able to assign role *{5}*.

In action *(17)*, the user condition R_u require role *{−1}* and role *{2}* to be able to assign role *{7}*.

If there's no other action to assign role *(5)* and role *(7)* and to revoke role *(1)* and role *(2)* than we can remove the action *(15)* since action *(15)* will never be executed.

3.1.6 Sort Actions

The MCMT module process actions in order, starting from the top to the bottom (from the first action to the last actions). This heuristic will arrange the administrative actions that MCMT can easily execute at the beginning.

During automated checking by the MCMT module (in the ASASPSPACETIME model), this module will, in turn, check each administrative action in the input data to calculate the states that can reach the goal by executing that action. In many cases, the administrative actions that executing that action reach the Goal from the initial state is at the end of the administrative actions set. Therefore, it is necessary to design the heuristic to perform the prioritization of administrative actions so that administrative actions that solve the problem will be prioritized for early analysis, thus, making the process faster and more efficient.

The heuristic will arrange the administrative actions closest to the initial state on the top position. Then, if the administrative action can be executed from the initial state, the MCMT module can quickly return the result. The priority is as follow:

- The administrative actions and user conditions which are set with "true" will be on the top because when these actions are checked, they will satisfy the initial state without checking any other actions.
- Next, the priority is given to actions that have administrative conditions that are "true" and that the user conditions have roles that are negative (and vice versa) because these actions are easily satisfied by the initial state since, by default, no user is assigned a role, so the user role condition is easily satisfied.
- Next, other administrative actions will be sorted by the total number of positive roles in administrative conditions and user conditions since the more positive roles need to be checked, the more time required. Administrative actions with fewer positive roles will be prioritized

Example 14: We will run this heuristic with the reachability problem in Example 8:

When running through the initial set, we will check the roles of the administrator conditions and user conditions to start the sorting.

In the first step, the actions taken are:

3. *Can_Assign_UA(true, l4, ts3, {true}, l3, ts2, 4)*
5. *Can_Assign_UA(true, l2, ts2, {true}, l4, ts2, 6)*
7. *Can_Assign_UA(true, l2, ts1, {true}, l3, ts2, 9)*

Actions *(3)*, *(5)*, *(7)* are taken because the admin conditions are true and the user conditions are true.

In the second step, the actions are updated:

3. *Can_Assign_UA(true, l4, ts3, {true}, l3, ts2, 4)*
5. *Can_Assign_UA(true, l2, ts2, {true}, l4, ts2, 6)*
7. *Can_Assign_UA(true, l2, ts1, {true}, l3, ts2, 9)*
8. *Can_Add_RE(true, l2, ts1, {−1 −10}, l4, ts2, 6)*

Action *(8)* is added because of the administrator conditions are true and the user conditions are the negative role

In the third step, the actions are updated to:

3. *Can_Assign_UA(true, l4, ts3, {true}, l3, ts2, 4)*
5. *Can_Assign_UA(true, l2, ts2, {true}, l4, ts2, 6)*
7. *Can_Assign_UA(true, l2, ts1, {true}, l3, ts2, 9)*
8. *Can_Add_RE(true, l2, ts1, {−1 −10}, l4, ts2, 6)*
1. *Can_Assign_UA(6, l5, ts4, {true}, l1, ts4, 4)*
2. *Can_Assign_UA(6, l2, ts1, {true}, l3, ts2, 4)*
6. *Can_Assign_UA(6, l4, ts2, {true}, l2, ts1, 9)*

Actions *(1)*, *(2)*, *(6)* are added because the positive roles in the administrator conditions and the user conditions in total are one condition.

In the fourth step, the actions to be updated are:

3. Can_Assign_UA(true, l4, ts3, {true}, l3, ts2, 4)
5. Can_Assign_UA(true, l2, ts2, {true}, l4, ts2, 6)
7. Can_Assign_UA(true, l2, ts1, {true}, l3, ts2, 9)
8. Can_Add_RE(true, l2, ts1, {-1 -10}, l4, ts2, 6)
1. Can_Assign_UA(6, l5, ts4, {true}, l1, ts4, 4)
2. Can_Assign_UA(6, l2, ts1, {true}, l3, ts2, 4)
6. Can_Assign_UA(6, l4, ts2, {true}, l2, ts1, 9)
4. Can_Assign_UA(6, l4, ts2, {9}, l2, ts1, 1)
9. Can_Add_RE(10, l2, ts2, {-1 -10}, l2, ts2, 9)
10. Can_Add_RE(12, l2, ts2, {10}, l2, ts2, 11)

Since the all the actions has been updated, the sorting is complete and the system will start safety checking.

3.1.7 Separating Complex Goal

Usually, this reachability problem will contain the goal *(u; C; l, ts)* in which *C* is a set of roles (complex goal). Obviously, solving the problem in C with the complex goal will be harder than with a single goal. In many cases, solving the single goal will likely return the result for the complex goal.

Therefore, the heuristic for complex goal is based on the idea of dividing complex goal into many single goals and start checking on every single goals. After that, based on the results received from the system, the heuristic can conclude the result for the initial complex goal. There are two main cases:

Case 1: If the system checks every single goal and one of the results is unreachable (one goal is not reached) then we conclude that the complex goal is not reached (unreachable). The reason is that for a complex goal to be reachable, all the roles in the set *C* are required to be reachable. If only a single goal in *C* (goal) cannot be reached, we conclude that the complex goal is unreachable, as well.

Case 2: If all results of every single goals are reachable, we still cannot conclude that the goal is reachable because there's a chance that the set of actions that make single goal reachable might not intersect (and therefore, the complex goal might not reachable). In this case, we will test the system again with the complex goal to return the final result. However, because ASASPSPACETIME only supports the single goal, this heuristic will add one more administrative action to the actions sets of the reachability problem to transform the complex goal into a single goal. The added administrative action is structured as the admin condition is true, the user condition contains the single goal, the new goal needs to execute with the maximum number of roles plus one, and the locations and timeslots are as same as the original goal. After that, this set of administrative actions (including newly added actions and new goal) will be included in ASASPSPACETIME for checking. If the result is not reached then the system returns unreachable and vice versa.

For Example: The goal *(3, l2, ts1)* is considered a single goal.

In cases, a new reachability problem with a complex goal *({3, 4, 9}, l2, ts1)*. In this case, the heuristic would divide the reachability problem with the goal *(3, 4, 9), l2, ts1)* into 3 subsets of the reachability problem with single goal *(3, l2, ts1), (4, l2, ts1), (9, l2, ts1)*. After that, the heuristic will call ASASPSPACETIME to check these three problems. Based on the results of these problems, the technique will determine the outcome of the complex goal.

3.1.8 Applying All Heuristics

In this section, we describe the overall architecture of the ASASPSPACETIME system with all the heuristics mentioned above. The architecture of the model is depicted in Fig. 9. The system will receive input as the reachability issue of the ASTRBAC policies. Then, the complex Goal module will be called to process the complex goal (if exists). Next, the system will run Forward and Backward. In the following step, the system will get the intersection of administrative actions of the Forward and Backward heuristics. After that, the system will combine any possible administrative actions. Then, the system will remove any duality administrator actions. Finally, the system uses the sorting heuristic to prioritize administrative actions. The final administrative actions will be used for safety analyzing. This final set of administrative actions with other configurations of the reachability problem will be the input for the Translator module to transform into BSR-STS (BSR-State Transitions System) reachability problem. Then, this problem will be analyzed by the module BSR to return the result of whether the violation of security requirement exists or not.

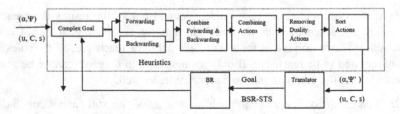

Fig. 9. ASASPSPACETIME with heuristics

It's also noted that the heuristics suggested in this research must ensure that the reachable/unreachable result is not altered with or without the use of heuristics. This is crucial because if the heuristics alter the result, for example, a goal in practice is reachable but using heuristics makes it become unreachable, the heuristics are meaningless as they do not find vulnerabilities in the system. Thus, deploying policies that are certified safe when using these heuristics will be very dangerous. The proof of the correctness of the heuristics will be based on the result of the backward reachability procedure (BR) as in the study of Ranise [15, 16].

4 Evaluations

All our experiments are performed on an Intel Core i7 CPU with 4 GB Ram running Ubuntu 12.04 LTS 32 bit. We use real scenario test cases, synthesized from [12, 14, 16, 19, 30], which contain university and hospital benchmark, and are widely used in security analysis communities. We run 5 experiments with our test cases. Our first experiments use the heuristics to run the experiment 15 times with different goals and calculate the average time for each test cases, the results are in Table 1.

Table 1. Our first experimental result with heuristics

#	Testcase	Config	Number of actions	Average runtime with heuristic (sec)
1	Test1	3 3 3	4	3.74
2	Test2	3 3 3	5	1.31
3	Test3	3 3 3	7	4.92
4	Test4	3 3 3	4	1.31
5	Test5	3 3 3	5	1.95
6	Test6	3 3 3	6	2.15
7	AGTHos1_test	16 4 5	125	2.06
8	AGTHos2_test	16 4 5	131	8.51
9	AGTHos3_test	35 6 10	165	1.01
10	AGTHos4_test	35 6 10	283	12.71
11	AGTHos5_test	16 8 20	355	141.79
12	AGTHos6_test	16 8 20	398	139.75
13	AGTUniv1_test	35 4 5	146	21.64
14	AGTUniv2_test	35 4 5	188	22.78
15	AGTUniv3_test	35 6 10	209	44.39
16	AGTUniv4_test	35 6 10	246	42.04

The first column shows our test case names; test cases 1 to 6 is our simple test cases created to test this program, test cases from 7 to 12 are taken from hospitals sets, test cases from 13 to 16 were taken from university sets. Each configuration contains 3 number representing max roles, max locations and max time slots in our STRBAC. The number of actions contain the number of administrative actions in ASTRBAC policies. We configure our experiment with a maximum number of roles, locations, and time slots.

In our second experiments, we use simple test cases, set 34 for the maximum number of roles, 10 for the maximum number of locations and 10 for the maximum number of time slots. Then, we keep increasing the number of actions and run these test cases with heuristics each time to get their average run time in Table 2 and Fig. 10.

Table 2. Our second experimental result with increasing number of actions

#	Testcase	Config	Number of actions	Average runtime (sec)
1	AGTHos1_a100	34 10 10	100	4.49
2	AGTHos1_a200	34 10 10	200	2.50
3	AGTHos1_a300	34 10 10	300	7.04
4	AGTHos1_a400	34 10 10	400	5.5
5	AGTHos1_a500	34 10 10	500	2.46
6	AGTHos1_a600	34 10 10	600	2.87
7	AGTHos1_a700	34 10 10	700	5.68
8	AGTHos1_a800	34 10 10	800	6.57
9	AGTHos1_a900	34 10 10	900	6.76
10	AGTHos1_a1000	34 10 10	1000	2.67

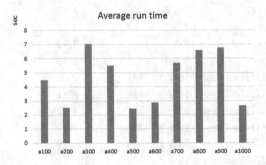

Fig. 10. Our average run time with increasing number of actions

In our third experiments, we use simple test cases without spatial-temporal constraints, set 16 for the maximum number of roles, and 10 for the maximum number of time slots. Then, we run these tests with randomly added locations each time to the test cases to get their average run time in Table 3 and Fig. 11.

Table 3. Our third experimental result with increasing number of locations

#	Testcase	Config	Number of actions	Average runtime (sec)
1	AGTHos1_l3	16 3 10	3	2.02
2	AGTHos1_l6	16 6 10	6	2.2
3	AGTHos1_l9	16 9 10	9	2.21
4	AGTHos1_l12	16 12 10	12	2.48
5	AGTHos1_l15	16 15 10	15	2.64
6	AGTHos1_l18	16 18 10	18	2.77

Fig. 11. Our average run time with increasing number of locations

In our fourth experiments, we try another test case, set 16 for the maximum number of roles, and 10 for the maximum number of locations. Then, we add random time slots each time to the test cases to get their average run time in Table 4 and Fig. 12.

Table 4. Our fourth experimental result with increasing number of time slots.

#	Testcase	Config	Number of actions	Average runtime (sec)
1	AGTHos1_t3	16 10 3	3	1.53
2	AGTHos1_t6	16 10 6	6	1.89
3	AGTHos1_t9	16 10 9	9	2.25
4	AGTHos1_t12	16 10 12	12	2.54
5	AGTHos1_t15	16 10 15	15	2.8
6	AGTHos1_t18	16 10 18	18	3.11

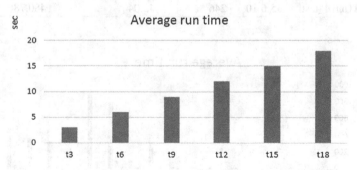

Fig. 12. Our average run time with increasing number of time slots

In our last experiments, we disable the heuristics to run the experiment 1 again 15 times with different goals and calculate the average time for each test cases, the results are in Table 5 and Fig. 13.

In our first and last experiments (Tables 1 and 5), we can conclude that our heuristics work and scalable with different goals. In our second experiments (Table 2), as the number of actions in ASTRBAC keeps increasing, the runtime of the heuristics is not affected much. The third and fourth experiments (Tables 3 and 4) clearly show that the increase in the number of time slots and the locations affect slightly to the run time of our system. After these experiments, we can conclude that our heuristics work and scalable, and our technique help reduce the runtime in the system when the number of actions, locations and the time slots keep increasing.

Table 5. Our last experimental result with and without heuristics.

#	Testcase	Config	Number of actions	Average runtime with heuristic (sec)	Average runtime without heuristic (sec)
1	Test1	3 3 3	4	3.74	7.58
2	Test2	3 3 3	5	1.31	1.53
3	Test3	3 3 3	7	4.92	8.4
4	Test4	3 3 3	4	1.31	2.48
5	Test5	3 3 3	5	1.95	4.05
6	Test6	3 3 3	6	2.15	13.14
7	AGTHos1_test	16 4 5	125	2.06	16.34
8	AGTHos2_test	16 4 5	131	8.51	87.67
9	AGTHos3_test	35 6 10	165	1.01	370.27
10	AGTHos4_test	35 6 10	283	12.71	106.29
11	AGTHos5_test	16 8 20	355	141.79	794.58
12	AGTHos6_test	16 8 20	398	139.75	913.67
13	AGTUniv1_test	35 4 5	146	21.64	251.47
14	AGTUniv2_test	35 4 5	188	22.78	317.48
15	AGTUniv3_test	35 6 10	209	44.39	367.51
16	AGTUniv4_test	35 6 10	246	42.04	480.78

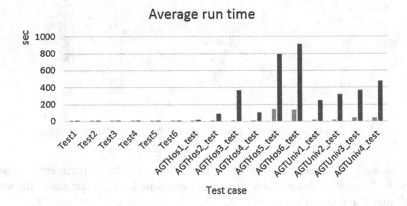

Fig. 13. Our average run time with and without heuristics

5 Conclusions

In this research, we introduce solutions to analyze the security policies of ASTRBAC with no prior knowledge of the number of users. Such solutions assist security designers in identifying potential security risks so that they can modify the policies to meet security requirements. If the policies have security holes which are deployed without modification, these policies may cause serious security problems consequently. Not only that, in cases of any violation of the initial security requirements, our solutions can also return the group actions, which by running these actions, the security requirement of the system may get into the security violation states. Our solutions can inform the security officers of the violation with useful information so that they can make appropriate decisions. We also focus on reducing the analysis runtime and improving performance by using heuristics that try to alleviate the state space problem.

An interesting line of research for future work is to consider role hierarchies in the analysis. Indeed, the presence of role hierarchies in the authorization policies will make the analysis much more complex because of its affect such as inheritance. Thus, a more sophisticated techniques processing such role hierarchies before the analysis should be designed and integrated into the proposed analysis technique.

Acknowledgement. This research is funded by Vietnam National University HoChiMinh City (VNU-HCM) under grant number **C2018-20-10**.

References

1. Samarati, P., de Vimercati, S.C.: Access control: policies, models, and mechanisms. In: Focardi, R., Gorrieri, R. (eds.) FOSAD 2000. LNCS, vol. 2171, pp. 137–196. Springer, Heidelberg (2001). https://doi.org/10.1007/3-540-45608-2_3
2. National Computer Security Center (NCSC): A guide to understanding discretionary access control in trusted system, Report NSCD-TG-003 Version1, 30 September 1987
3. Osborn, S.: Mandatory access control and role-based access control revisited. In: Proceedings of the 2nd ACM Workshop on Role-Based Access Control, RBAC 1997, pp 31–40. ACM (1997)
4. Sandhu, R., Coyne, E.J., Feinstein, H.L., Youman, C.E.: Role-based access control models. IEEE Comput. **29**, 38–47 (1996)
5. Ferraiolo, D., Kuhn, R.: Role-based access control. In: 15th NIST-NCSC National Computer Security Conference, pp. 554–563, October 1992
6. Sandhu, R., Ferraiolo, D., Kuhn, R.: The NIST model for role-based access control: toward a unified standard. In: 5th ACM Workshop Role-Based Access Control, pp. 47–63, July 2000
7. Sandhu, R., Bhamidipati, V., Munawer, Q.: The ARBAC97 model for role-based administration of roles. ACM Trans. Inf. Syst. Secur. (TISSEC) **2**, 105–135 (1999)
8. Kumar, M., Newman, R.: STRBAC - an approach towards spatiotemporal role-based access control. In: Proceedings of the Third IASTED International Conference on Communication Network and Information Security CNIS, pp. 150–155 (2006)
9. Sharma, M., Sural, S., Atluri, V., Vaidya, J.: An administrative model for spatio-temporal role based access control. In: Bagchi, A., Ray, I. (eds.) ICISS 2013. LNCS, vol. 8303, pp. 375–389. Springer, Heidelberg (2013). https://doi.org/10.1007/978-3-642-45204-8_28

10. Li, N., Tripunitara, M.: Security analysis in role-based access control. In: The Proceedings of ACM Symposium on Access Control Models and Technologies, pp. 126–135. ACM Press (2004)
11. Jha, S., Li, N., Tripunitara, M., Wang, Q., Winsborough, H.: Towards formal verification of role-based access control policies. IEEE TDSC **5**(4), 242–255 (2008)
12. Gofman, M.I., Luo, R., Solomon, Ayla C., Zhang, Y., Yang, P., Stoller, S.D.: RBAC-PAT: a policy analysis tool for role based access control. In: Kowalewski, S., Philippou, A. (eds.) TACAS 2009. LNCS, vol. 5505, pp. 46–49. Springer, Heidelberg (2009). https://doi.org/10.1007/978-3-642-00768-2_4
13. Jayaraman, K., Tripunitara, M., Ganesh, V., Rinard, M., Chapin, S.: Mohawk abstraction-refinement and bound-estimation for verifying access control policies. ACM TISSEC **15**, 18 (2013)
14. Ferrara, A.L., Madhusudan, P., Nguyen, T.L., Parlato, G.: VAC - verifier of administrative role-based access control policies. In: Biere, A., Bloem, R. (eds.) CAV 2014. LNCS, vol. 8559, pp. 184–191. Springer, Cham (2014). https://doi.org/10.1007/978-3-319-08867-9_12
15. Ranise, S., Truong, A., Vigano, L.: Automated analysis of RBAC policies with temporal constraints and static role hierarchies. In: the Proceeding of the 30th ACM Symposium on Applied Computing (SAC15), pp. 2177–2184. ACM (2015)
16. Ranise, S., Truong, A., Armando, A.: Scalable and precise automated analysis of administrative temporal role-based access control. In: Proceedings of the 19th ACM Symposium on Access Control Models and Technologies, pp. 103–114. ACM (2014)
17. Truong, A., Ranise, S.: ASASPXL: new clother for analysing ARBAC policies. In: Dang, T.K., Wagner, R., Küng, J., Thoai, N., Takizawa, M., Neuhold, E. (eds.) FDSE 2016. LNCS, vol. 10018, pp. 267–284. Springer, Cham (2016). https://doi.org/10.1007/978-3-319-48057-2_19
18. Ghilardi, S., Ranise, S.: MCMT: a model checker modulo theories. In: Giesl, J., Hähnle, R. (eds.) IJCAR 2010. LNCS (LNAI), vol. 6173, pp. 22–29. Springer, Heidelberg (2010). https://doi.org/10.1007/978-3-642-14203-1_3
19. Harrison, M., Ruzzo, W., Ullman, J.: Protection in operating systems. Commun. ACM **19**, 461–471 (1976)
20. Bertino, E., Bonatti, P., Ferrari, E.: TRBAC a temporal role-based access control model. ACM TISSEC **4**(3), 191–233 (2001)
21. Joshi, J., Bertino, E., Latif, U., Ghafoor, A.: A generalized temporal role-based access control model. IEEE Trans. Knowl. Data Eng. **17**, 4–23 (2005)
22. Kumar, M., Newman, R.: STRBAC - an approach towards spatio-temporal role-based access control. In: Communication, Network, and Information Security, pp. 150–155 (2006)
23. Aich, S., Mondal, S., Sural, S., Majumdar, A.K.: Role based access control with spatiotemporal context for mobile applications. In: Gavrilova, M.L., Tan, C.J.K., Moreno, E.D. (eds.) Transactions on Computational Science IV. LNCS, vol. 5430, pp. 177–199. Springer, Heidelberg (2009). https://doi.org/10.1007/978-3-642-01004-0_10
24. Uzun, E., Atluri, V., Sural, S., Vaidya, J., Parlato, G., Ferrara, A.: Analyzing temporal role-based access control models. In: SACMAT, pp. 177–186. ACM (2012)
25. Ghilardi, S., Ranise, S.: Backward reachability of array-based systems by SMT solving termination and invariant synthesis. Log. Methods Comput. Sci. **6**, 1–48 (2010)
26. http://research.microsoft.com/en-us/um/redmond/projects/z3
27. Ranise, S.: Symbolic backward reachability with effectively propositional logic. Appl. Secur. Policy Anal. FMSD **42**, 24–45 (2013)
28. Piskac, R., Moura, L., Bjørner, N.: Deciding effectively propositional logic using DPLL and substitution sets. J. Autom. Reason. **44**, 401–424 (2010)

29. Sasturkar, A., Yang, A., Stoller, S., Ramakrishnan, C.: Policy analysis for administrative role-based access control. In: 19th IEEE Computer Security Foundations Workshop, pp. 124–138 (2006)

30. Dinh, K.K.Q., Tran, T.D., Truong, A.: Security analysis of administrative role-based access control policies with contextual information. In: Dang, T.K., Wagner, R., Küng, J., Thoai, N., Takizawa, M., Neuhold, E.J. (eds.) FDSE 2017. LNCS, vol. 10646, pp. 243–261. Springer, Cham (2017). https://doi.org/10.1007/978-3-319-70004-5_17

Classification Methods in Colon Disease Information System

Anna Kasperczuk[✉] and Agnieszka Dardzinska

Department of Mechanical Engineering, Division of Biocybernetics
and Biomedical Engineering, Bialystok University of Technology,
ul. Wiejska 45c, 15-351 Bialystok, Poland
{a.kasperczukuk, a.dardzinska}@pb.edu.pl

Abstract. This paper presents the process of building a new logistic regression model, which aims to support the decision-making process in medical database. The developed logistic regression model, J48 classifier and Random Tree algorithm define the probability of the disease and indicates the statistically significant changes that affect the onset of the disease. In our work, we attempted to build a classifier that would classify patients undergoing ulcerative colitis and other conditions within the lower gastrointestinal tract. The value of probability can be treated as one of the feature in decision process of patient's future treatment.

Keywords: Selection · Classification · Decision system · Information system

1 Introduction

Rapid development of computerization gave rise to the possibility of quick access to information sources and storage of huge data resources. Technological progress has made it necessary to look for methods that have made it possible to analyze and process information. There was a need to develop new methods of data analysis. The methods developed in the context of knowledge mining from databases proved to be the answer to this demand. A number of methods have been created that fit in two main areas of data mining: statistics and data mining. Currently, data analysis opens up opportunities for extracting medical knowledge. This is extremely important, considering the search for symptoms and various factors conditioning the emergence of diseases or enabling the distinction of disease entities. The use of data analysis methods also leads to the creation of knowledge bases, in which the collected data can be used in further knowledge mining processes, indicating specific features that should be changed, e.g. in the treatment process, to improve the patient's health. This fact is testimony about the great possibilities offered by data analysis using statistical methods or data mining methods. In this paper, the factors causing the onset of intestinal disease during experiments are sought. Ulcerative colitis is a disease that causes long-term inflammation of the colon, which creates irritation or ulcers. This can lead to debilitating abdominal pain and potentially life-threatening complications. It affects only the colon or rectum and destroys the innermost part of the mucosa, not passing through the mouth. Ulcerative colitis causes inflammation and ulcers in the large intestine, which can cause a frequent feeling of need for bowel movement. Exact causes of the disease are not known, therefore their search is extremely important.

Work in an orderly manner introduces issues related to the extraction of knowledge from information systems. The second section is devoted to the description of the information system definition. The next third section discusses the issue of classification. Discussed are decision tree algorithms, which in the medical expression were used to build classifiers. The problem of logistic regression has been discussed extensively. Logistic regression models are used to explain the relationship between independent variables and dependent variable expressed on the dichotomic scale. Subsequent parts of the work are devoted to the description of the research sample, the experiment and the results of the research. The last part was devoted to the conclusions.

2 Main Assumptions

We take into consideration $S = (X, A, V)$, which is an information system, where:

- X is a nonempty, finite set of objects,
- A is a nonempty, finite set of attributes,
- $V = \{V_a : a \in A\}$ is a set of all attributes values.

Additionally, $a : X \rightarrow V_a$ is a function for any $a \in A$, that returns the value of the attribute of a given object [4]. The attributes are divided into different categories: set of stable attributes A_{St}, set of flexible attributes A_{Fl} and set of decision attributes D, such that $A = A_{St} \cup A_{Fl} \cup D$. In this paper we analyze information systems with only one decision attribute d. The example of an information system S is represented as Table 1 [7, 15].

Table 1. Information system S

X	a	b	c	d
x_1	a_1	b_2	c_2	d_1
x_2	a_1	b_1	c_1	d_1
x_3	a_2	b_1	c_1	d_1
x_4	a_2	b_2	c_1	d_2
x_5	a_2	b_2	c_2	d_2
x_6	a_2	b_1	c_1	d_1
x_7	a_2	b_2	c_1	d_2
x_8	a_2	b_1	c_2	d_2

Information system is represented by eight objects, one stable attribute a, two flexible attributes b, c and one decision attribute d.

In many cases, decision-making processes are based on multiple regression models, i.e. in which we analyze the influence of several independent variables on one dependent variable of the measurable type [16].

In contrast, when a dependent variable is of a dichotomous type, we can apply logistic regression. In economic studies, a very popular example of the application of logistic regression is the analysis of the ability to repay bank loans, while in medicine the possibility of indicating the probability of, for example, the occurrence of a certain disease, from the point of view of the (statistically significant) characteristics of the patient and the specificity of the disease. The advantage of logistic regression is that the interpretation of results is very similar to methods used in classical regression method. However, it should also be noted that compared to multiple regression, logistic regression is more complex computationally [1].

3 Classification

The classifier is an algorithm that implements classification, especially in a concrete implementation. We use for this classification - model finding process that is used to partitioning data into different classes according to some constrains. In other words, we can say that classification is the process of generalizing the data according to different instances. There are many different classifiers and many different types of dataset resulting in difficulty in knowing which will perform most effectively in any given case. It is already widely known that some classifiers perform better than others on different datasets. It is always possible that another classifier may work better. In deciding which classifier will work best for a given dataset there are two options. First is to put all the trust in an expert's opinion based on knowledge and experience. Second is to run through every possible classifier that could work on the dataset, identifying rationally the one which performs best [5, 7].

Classification is a data mining algorithm that creates a step-by-step guide for how to determine the output of a new data instance. It is the process of finding a set of models that differentiate data classes and concepts. We used it to predict group memberships for data instances. In first step we describe a set of predetermined classes. Each tuple is assumed to belong to a predefined class as determined by class label attribute, the set of tuples are used for model construction, called training sets. The model is represented as classification rules, decision trees or mathematical formulas. This model is used to classifying future data trends and unknown objects. It estimates the accuracy of the constructed model by using certain test cases. Test sets are always independent of the training sets [7, 8].

3.1 Decision Trees

Among the classification methods, one of the most popular method is is the induction of decision trees. It is particularly attractive because of the intuitive way of knowledge representation understood by people [20].

Initially decision trees appeared in the 1960s in the areas of research on psychology and sociology. In informatics, for the first time they found their application in the works in the 80 [2, 21].

Compared to other methods of classification, decision trees can be constructed relatively quickly. Their main advantage is the clear representation of knowledge, the possibility of using multidimensional data, and scalability with the use of large data sets. In addition, the accuracy of this method is comparable to the accuracy of other classification methods. However, the main disadvantage of the discussed method is the high sensitivity to the missing values of attributes, because at their bases there is an explicitly expressed assumption of full availability of information gathered in the database. The disadvantages also include the inability to capture the correlation between attributes [21].

In the decision tree method, the result of their action is a directed, consistent graph with a tree structure. He uses the graphical data structure and presents their possible consequences, helping in making decisions. The structure obtained in this way is a set of decision nodes connected by means of "branches" that propagate down from the "root" to the ending "leaf" tree.

Classification trees are used to determine the affiliation of objects to the quality class of a dependent variable. This is done based on measurements of one or more prediction variables. The classification tree presents the process of dividing the set of objects into homogeneous classes. The division is based on the values of the features of the objects, the leaves correspond to the classes to which the objects belong, while the edges of the tree represent the values of the features on the basis of which the division was made [21].

The process of creating a decision tree is based on the recursive division of the teaching set into subsets, which takes place to achieve their homogeneity due to the belonging of objects to classes. The goal is to create a tree with the fewest number of nodes, and as a consequence, the simplest classification rules [2].

The decision tree creation algorithm can be written as follows [10, 18]:

- For a given set of objects it should be checked whether they belong to the same class (if they belong - end the proceedings, if they do not belong - consider all possible divisions of a given set into the possibly homogeneous subsets);
- Evaluate the quality of each of these subsets according to the previously accepted criterion and select the best one;
- Split the set of objects in the chosen way;
- Perform steps for each of the subsets.

ID3 algorithm

One of the earliest proposals for implementing learning systems and acquiring knowledge presented in the form of a decision tree is the ID3 algorithm, which was developed by Quinlan [21]. It generates a decision tree based on a series of unit cases. The decision tree is a structural record of the knowledge extracted, allowing on the basis of the values of certain features to assign specific values to decision-making [7, 21].

In order to generate a decision tree with the ID3 algorithm, a relatively large set of examples describing a given situation is necessary. Each example from the set takes a specific value for each attribute from the list of conditional attributes and the decision attribute. Each attribute describing a given example takes one value from the list of possible values. The described set of examples is a training set. When a list of attributes is given together with lists of available values and a training set, it is possible to start building a decision tree [18].

The ID3 algorithm:
Input: training set D, set of conditional attributes A, method of selecting the partition point SS.
Output: decision tree rooted at the apex N.
procedure BuildTree (D, A, SS):
create the top of the decision tree N;

if all records of set D belong to the same class C then

return the vertex N as the leaf of the decision tree and assign the class C label to the given vertex;

end if

if attribute_list A is empty then

return the vertex N as the number of the tree and assign the label of the dominant class to the vertex in the training set D;

end if

use the SS method to select the subdivision attribute from file A;
assign the attribute tip to the vertex of N;
for all values of the a_i *attribute-division* to
$S_i \leftarrow$ set of D records with the attribute value-division = a_i;
$N_i \leftarrow$ *BuildTree* (S_i, (attribute list A) - (*attribute-division*), SS);
create an edge with N to Ni labeled with the value of ai;

end for

return tip N.

The disadvantage of the ID3 algorithm is operation only for nominal attributes, with incomplete data the algorithm does not work. In addition, the final trees are too adapted to the data and the measure of information gain favors features with a large number of values. The described algorithm is characterized by the lack of resistance to the phenomenon of overfitting, because it does not cope with data that disturbs their general information, which can lead to a high error rate on test data [10, 18]. These problems were eliminated after the introduction of subsequent versions of the ID3 algorithm (including C4.5).

C4.5 algorithm

The C4.5 algorithm is one of the two most popular algorithms used in practice. This algorithm is in fact an extension of the ID3 algorithm.

The C4.5 algorithm recursively passes through all nodes, selecting the possible division until further subdivisions are possible. For qualitative attributes, this algorithm by definition creates separate branches for each value of the qualitative attribute. This may be the result of a greater branching of the tree than is desirable, because some values may be rare and be naturally related to other values [3, 10].

For continuous attributes, the discussed algorithm considers all possible subdivisions into two subsets, determined by the division point w. Unlike discrete attributes, continuous attributes can appear on multiple levels of the same branch of the decision tree. For each of the possible divisions, its quality is assessed by measuring the relative value of the information gain. Selects the option that gives maximum information gain [10, 18].

The C4.5 algorithm, in addition to the decision tree induction method, makes it possible to transform a tree into a set of rules. The rules are treated here as a classification model different from the tree, because they are not a faithful representation of the tree.

3.2 Regression Analysis

The object of regression analysis belongs to the applications of mathematics, related to the solution of some problems of mathematical statistics using probabilistic methods. The primary goal of this field is to determine the relationship between the quantities we are interested in (variables), its character and strength, as well as the structure of the model that describes this relationship well. Most often, the only way to study such relationships is to conduct experiments and only in a few cases they can be obtained theoretically. Examples of such dependencies are: dependence of soil yield on various mineral fertilizers; dependence of the bank's profit level on the number of clients, the amount of investment, the amount of loans issued, etc.

It is worth noting that the word regression in Latin translation means undoing. The use of the word regression in the name of this domain, as well as in the name of several other terms, is historical and rather unfortunate. This name was probably first used in 1885 by the English scientist Sir F. Galton (Darwin's student) while studying the relationship between the growth of offspring and parental growth. He showed that extremely tall parents (much higher above mediocrity) have children of lower growth, while parents with a significantly lower than average rise have children higher than them. Galton called this phenomenon a backwardness towards mediocrity. But in fact, the field dealing with the search for dependencies on the basis of experimentation is much older: for example, French mathematicians (especially Laplace) in the eighteenth century carried out analyzes that we would call regression.

Let, then, let us be interested in quantities (hereafter we call them variables) $Y, X^{(1)}, \ldots, X^{(m)}, m \geq 1$. The variable Y is called the dependent variable and the variables $X^{(1)}, \ldots, X^{(m)}$, independent variables. The question we would like to get an answer to is: is there a relationship between Y and $X^{(1)}, \ldots, X^{(m)}$, (in other words, whether the variables $X^{(1)}, \ldots, X^{(m)}$, affect Y)? And if the dependence is, then we would like to express it with a certain model (equation) [6].

Let us assume that we can carry out n experiments, n measurements of the variable Y size depending on certain sizes of variables $X^{(1)}, \ldots, X^{(m)}$,. The values of the above variables, obtained during the experiments, will mean the appropriate lower-case letters. So, the starting point in our reasoning will be observations $\left(x_i^{(1)}, \ldots, x_i^{(m)}, y_i \right)$, $i = 1, \ldots, n$.

It is worth noting that variables can be related to each other by functional dependence or statistical dependence. The functional relationship is characterized by the fact that each one of the values of independent variables is represented by only one, univocally determined value of the dependent variable (for example, the square field is a function of its side). However, we rarely deal with data that accurately describes a similar relationship. More often we deal with the so-called statistical dependence. The statistical relationship is based on the fact that specific values of independent variables correspond to strictly defined mean values of the dependent variable. This dependence can be described using the function $E\left(Y|X^{(1)} = x^{(1)}, \ldots, X^{(m)} = x^{(m)}\right)$.

Well, we assume further that the values of the dependent variable are random (the values of independent variables will conveniently be considered non-random). So, the observed values $\{y_i\}$ contain some random components, often called or interpreted as measurement errors. In fact, the sources of these random factors can be very different: from errors caused by the imprecision of measuring devices to errors caused by not taking into account all factors affecting the dependent variable [6, 16].

As an illustration of the concept of statistical dependence, consider the following example, taken from [17].

Example 1
Let us consider the results of the colloquium (scale from 0 to 25 points) and the final exam (scale from 0 to 50 points) from the mathematical statistics. 19 students of a technical school took part in the colloquium and the exam. The results of the colloquium and the exam are given in the Table 2:

Table 2. The results for example 1

Number	1	2	3	4	5	6	7	8	9	10
Test	7	11	12	14	17	15	21	22	19	13
Exam	20	24	25	30	35	30	43	42	41	24
Number	11	12	13	14	15	16	17	18	19	
Test	5	12	16	14	21	20	17	10	17	
Exam	14	27	35	28	42	40	34	23	40	

The relationship between the result of the final exam (dependent variable) and the colloquium (independent variable) is presented in Fig. 1. This figure is called the scatterplot and is a useful graphical representation of dependencies between variables. It is created by pairs of points $(x_i, y_i), i = 1, \ldots, 19$. Let us note that in the case when we deal with one independent variable, the problem of model construction should always start with the preparation of a scatter chart. We see that we are dealing here with a statistical dependency, not a functional one, because we have students whose result the colloquium is the same, but the result of the exam differs (for example, students with numbers 5, 17 and 19 have the same test result – 17 points, but different result of the exam – 35, 34 and 40 points respectively).

Generally, the process of model construction proceeds in the following stages.

Stage 1. Model specification. At this stage, we make a choice of the model's form (linear, square, nonlinear, etc.) which we will consider. The choice is made on the basis of a scatter chart, or some knowledge about the possible nature of dependence.

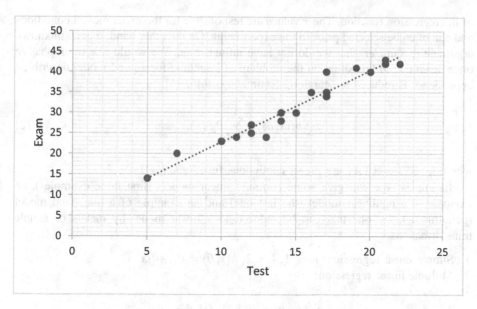

Fig. 1. Scatter chart

We can also suggest searching for the simplest solution, i.e. choosing a linear model.

Stage 2. Statistical inference about the parameters of the model. At this stage, using appropriate statistical methods and based on the data we have, we estimate the parameters of the model and, if necessary, test hypotheses about these parameters.

Stage 3. Model verification. Next, it should be checked whether the model constructed at the previous stage under certain assumptions actually meets these assumptions and whether it fits well with the possessed data. Here, too, using statistical methods, we estimate the significance of the parameters obtained. If the model does not meet the requirements, we formulate a new model and go back to the previous stage.

Stage 4. Using the model. If the created model is considered correct, then we can use it further, for example for predicting the value of a dependent variable in the case of other than the values obtained so far, the values of independent variables, or for controlling - that is, determining the values of independent variables to obtain the appropriate value of the dependent variable.

Thus, according to stage 1 (stage of model selection), we narrow the circle of the considered functions describing dependence to a certain parametric class of the H function, i.e. we assume that the model is described by a function from class H:

$$H = \left\{ h(x, \theta), \theta \in \Theta \subset \mathbb{R}^k, x = \left(x^{(1)}, \ldots, x^{(m)} \right) \in \mathbb{R}^m \right\}, \tag{1}$$

where $h : \mathbb{R}^m \times \mathbb{R}^k \to \mathbb{R}$ is a continuous function. The function h is called the regression function. It is worth emphasizing the different role of the x and θ arguments of the regression function. The x values are responsible for the experimental conditions and can often be selected prior to the experiments. On the other hand, θ is an unknown argument (parameter) that we do not have influence on. We would like to emphasize once again that in this situation the problem of searching for a good model describing dependence is reduced to searching (estimating) θ [6, 16]

$$h((x, \theta)) = \sum_{j=1}^{k} \theta_j f_j(x),\tag{2}$$

where $f_j : \mathbb{R}^m \to \mathbb{R}$ they are preset continuous functions, $j = 1, \ldots, k$.

Examples: We will give some common linear models (first three examples), an example of a nonlinear model (the last one) and an example of a linearized model (penultimate), i.e. one that can be reduced to a linear model by means of simple transformations.

1. Simple linear regression: $m = 1$, $k = 2$, $h(x, \theta) = \theta_0 + \theta_1 x$.
2. Multiple linear regression:

$$k = m + 1, h(x, \theta) = \theta_0 + \theta_1 x^{(1)} + \ldots + \theta_m x^{(m)}.$$

3. Polynomial regression degree $p > 1$:

$$m = 1, k = p + 1, h(x, \theta) = \theta_0 + \theta_1 x + \ldots + \theta_p x^p.$$

4. Power regression: $m = 1$, $k = 2$, $h(x, \theta) = \theta_0 x^{\theta_1}$. In this case, we use the logarithmic function to import the regression function to linear form: $h'(x', \theta') = \theta_0' + \theta_1' x'$, where $h' = \ln h, \theta_1' = \ln \theta_0, \theta_1' = \theta_1, x' = \ln x$.
5. Nonlinear regression: $m = 1$, $k = 3$, $h(x, \theta) = \theta_0 + \theta_1 e^{-\theta_2 x}$. In this situation, no transformation will bring a regression function to the linear function of parameters.

3.3 Logistic Regression Models

Logistic regression models are used to explain qualitative variables depending on the level of exogenous variables (qualitative or quantitative). Logistic regression has important applications, for example in modeling the risk of finding themselves in a certain unit testing condition. We are dealing with a binomial model if the explanatory variable accepts two states, i.e. whether the investigated phenomenon occurs or not.

Nowadays, logistic models are widely used in banks to assess credit risk and enterprises to assess customer loyalty. They are also one of the tools used by actuaries to assess insurance risk to evaluate the chances of conversion and retention of insurance policies [7]. In life insurance, this model allows estimating the probability of death based on the underlying demographic characteristics such as sex, age, place of residence [6, 9, 11], and also if we have a sufficiently large history database of insured clients, we can include information collected through medical surveys.

In the field of demography, parametric (analytical) models of the human survival process (so-called measurability law) are usually sought only in dependence from age, building models for men and women separately (possibly for people living in cities and in the countryside). For this purpose, curvilinear regression is used. As analytical models of death rates, the following functions are most often used: exponential, power, polynomial, polynomial-exponential, and logistic functions [18]. Access to more and better-quality databases and the development of numerical methods and computer software make these models more and more complex. There are many attempts to create demographic analytical models in the literature, but rarely meet the use of generalized linear models (GLM), including logistic regression, for the analysis of demographic data.

It is especially useful when the dependent variable is a qualitative variable. Here we will limit ourselves to considering the case when it only accepts two values: 1 and 0 (example: 1 - the event will occur, 0 - the event will not occur). In this situation, the use of linear regression is unhelpful and may even be devoid of an interpretative sense.

The logistic regression model for the dichotomous variable can be expressed by the formula [13]:

$$Y \sim B(1,p),$$
$$p = E(Y|X) = \frac{\exp(\beta X)}{1 + \exp(\beta X)}. \tag{3}$$

Where $B(1,p)$ is a binomial distribution with the probability of success p.

Last equality assumes the choice of the canonical join function - logit. Modeling p with logit allows for a convenient interpretation of logistic regression results in terms of chance.

The odds are a function of probability. Instead of calculating the classical probability, i.e. the ratio of the number of successes to the number of all trials, we calculate the ratio of the probability of the success to the probability of the failure. Let's o mean a chance and a probability of success [16, 19]. Then:

$$o = \frac{p}{1-p} \tag{4}$$

$$p = \frac{o}{1+o} \tag{5}$$

The probability of the event $p \in (0,1)$, so the chance takes values in the range $(0,\infty)$, and its logarithm of the range $(-\infty,\infty)$.

Example 2

Let us consider hypothetical events: A - in a sample of 100 people smoked cigarettes 80 got cancer and B - out of 100 non-smokers fell ill 10.

Then $o(A) = \frac{80}{20} = 4, o(B) = \frac{10}{90} \approx 0.1$. This means that the probability of occurrence of event A is fourth times greater than the probability of its occurrence among smokers. We can also say that the probability of the event A is 4: 1 (similarly interpret the event B).

Logistic regression is based precisely on such way of expressing probability [9]. In the logistic regression model for one explanatory variable X_1 the chance is equal to:

$$\frac{P(X)}{1 - P(X)} = \exp(\beta_0 + \beta_1 X_1) \tag{6}$$

While the logarithm of chance:

$$\log\frac{P(X)}{1 - P(X)} = \beta_0 + \beta_1 X_1 \tag{7}$$

The logarithm of opportunity is linearly dependent on the explanatory variable X_1, so that β_1 can be easily interpreted. This factor tells us about the change in the logarithm of the opportunity associated with the change in unit of the factor described by X_1. By going from the logarithm of chance to chance of β_1 is a relative change in the occurrence of an event by the factor described by the variable X_1.

- If $e^{\beta_1} > 1$ the factor described by the variable X_1 has a stimulating effect on the occurrence of the studied phenomenon.
- If $e^{\beta_1} < 1$ the factor is limiting.
- If $e^{\beta_1} = 1$ this factor does not affect the event described.

The odds ratio is used for comparison of two classes of observations. This is the odds ratio that the event occurs in the first group elements, and that it also occurs in the other. It is described by the formula [19]:

$$OR = \frac{p_1}{1 - p_1}\frac{1 - p_2}{p_2} = \frac{p_1(1 - p_2)}{p_2(1 - p_1)} \tag{8}$$

Where p_i means the probability of an event in the i-th class of observation. We interpret them as follows:

- If $OR > 1$, then in the first group the occurrence of the event is more likely.
- If $OR < 1$, the event occurs in the second group more likely.
- If $OR = 1$, then the event is equally likely in both observational classes.

Example 3
Based on the data from the previous example, we calculate the odds ratio of group A to group B. We thus: $OR = \frac{4}{0.1} = 40$. The chance of developing cancer in smokers is 40 times higher than that of non-smokers.

The logistic regression model does not require some of the assumptions needed for linear regression. Vector explanatory variables and the rest need not have normal distribution, heteroskedasticity is acceptable. However, it is necessary to meet several other conditions [9]:

- The relationship between the log of opportunity and the vector of explanatory variables must be linear.
- The explanatory variable must be binary, where the level coded as "1" represents the desired result (success).
- Observations must be independent - we use this to derive the form of the reliability function.
- The model must be well matched, that is to say only those explanatory variables that affect the explanatory variable, and do not ignore any such variable.
- There is no strong collinearity in data - it is a source of numerical problems.

The last two conditions are more of a guideline than assumption. We do not use them to derive logistic regression theory, but a model that does not meet them can lead to incorrect conclusions.

Likelihood Function
We derive the form of the reliability function L for logistic regression. The explanatory variable Y is binary and for single observation i and occurs:

$$Y_i|X_i = \begin{cases} 1 \; with \, probability \, p(X_i) \\ 0 \; with \, probability \, 1 - p(X_i) \end{cases} \qquad (9)$$

Hence:

$$L(X_i, \beta) = P(Y_i = 1|X_i)^{Y_i} \cdot P(Y_i = 0|X_i)^{1-Y_i} = p(X_i)^{Y_i}[p(X_i)]^{1-Y_i} \qquad (10)$$

Wherein the vector of estimated parameters β is involved as a function of p, in accordance with the formula:

$$L(X_i, \dots, X_n, \beta) = \prod_{i=1}^{n} p(X_i)^{Y_i}[1 - p(X_i)]^{1-Y_i} \qquad (11)$$

The likelihood function is used to estimate the parameters β and the maximum likelihood for the hypothesis testing.

Testing of Hypothesis
In particular, the hypotheses of the statistical significance of variables (the hypothesis on the question of whether a model that contains a certain variable) provides much more information about the variable elucidated from the model without that variable. Testing such hypothesis is based on comparing the observed values of the explanatory variable Y with its $\hat{Y}$ matched values by two models, one with the explanatory variable we are interested in.

We introduce the concept of saturated model. It concerns not only logistic regression, but the whole class of generalized linear models [6].

By saturated model we mean a model with a number of parameters equal to the number of observations.

For example, for a set of data with two observations the full model is such as: $g(EY) = \beta_0 + \beta_1 X_1$. The concept of a full model enables another interpretation of the value of the observed variable being interpreted - as matched values of a full model for a given dataset.

The significance test of the explanatory variable uses deviation statistics D:

$$D = -2\log\left[\frac{value\ of\ the\ likelihood\ function\ of\ the\ estimation\ model}{value\ of\ the\ likeluhood\ function\ of\ the\ full\ model}\right] \qquad (12)$$

Multiplied by the (-2) logarithm of the likelihood ratio has a known distribution, so it is suitable for testing statistical hypotheses. The tests that are based on it are the reliability quotient tests:

$$D = -2\sum_{i=1}^{n}\left[Y_i \log\left(\frac{p(X_i)}{Y_i}\right) + (1 - Y_i)\log\left(\frac{1 - p(X_i)}{1 - Y_i}\right)\right] \qquad (13)$$

In the logistic regression model, the value of the reliability function for the full model is 1, which can be shown by inserting $p(X_i) = Y_i$ (property of the full model) into (13):

$$L(full\ model) = \prod_{i=1}^{n} Y_i^{Y_i}[1 - Y_i]^{1-Y_i} = 1 \qquad (14)$$

From (10) and (12) we get:

$$D = -2\log(value\ of\ the\ reliability\ function\ of\ the\ estimated\ model) \qquad (15)$$

In addition to the likelihood-ratio test, there are two alternative methods of testing the significance of the explanatory variables: Wald test and test Score [9]. As with the reliability quotient test, it is needed to set a sufficient number of observations n. The Wald test is divided by the estimation of the parameter at variable X_* by the standard error of this estimate (as determined by the SE):

$$W = \frac{\hat{\beta}_*}{SE\left(\hat{\beta}_*\right)} \qquad (16)$$

Based on null hypothesis $(\beta_* = 0)$, W has asymptotically decomposed $N(0, 1)$. The test score is based on the statistics obtained from the derivative of the logarithm of the likelihood function and does not require the calculation of the MNW estimators of the β parameters. The term score is determined derivative of the logarithm of the likelihood function:

$$U(\beta) = \frac{\partial logL(\beta|x)}{\partial \beta} \qquad (17)$$

The test statistic for the null hypothesis $\beta_* = 0$ has asymptotically χ^2 distribution and amounts:

$$S = U(0)^2 I(0)^{-1} \qquad (18)$$

Where $U(0)$ score of parameter β_* and I is the Fisher information.

Determination of Confidence Intervals

Confidence intervals for factor estimates in the logistic regression model are constructed on the basis of the Walt test statistics. Using that W has an asymptotic standard normal distribution [6]:

$$W = \frac{\hat{\beta}_*}{SE\left(\hat{\beta}_*\right)} \sim N(0,1) \qquad (19)$$

The limits of the confidence interval at significance level $(100 - \alpha)\%$ for estimating the parameter β are:

$$\hat{\beta} \pm z_{1-\frac{\alpha}{2}} SE\left(\hat{\beta}\right) \qquad (20)$$

Where $z_{1-\alpha/2}$ is the quantum decomposition $N(0,1)$ of order $1 - \alpha/2$.

In mathematical statistics, the ROC curve is a graphical representation of the predictive model efficiency by plotting the binary classifier's quality characteristics using multiple cutoff points. Other words - each point of the ROC curve corresponds to another error matrix obtained by modifying the cut-off point. The more different points we cut, the more points we get on the ROC curve. Finally, we put TPR (True-Positive Rate) and FPR (False-Positive Rate) [6, 22].

The ROC curve, as a function of the cutoff point, represents the TPR variance (coverage of the actual positive class coverage) depending on the FPR (level of error actually negative). There is a compromise when we choose "cut-off" and we want to maximize TPR "keeping in line" FPR error.

The additive parameter that is defined for the classifier which is the regression model is the AUC. The AUC (Area Under the ROC) interpretation is the probability that the predicted predictive model will score higher (score) a random element of a positive class from a random negative element [4].

4 Experiments

In the experiment, an attempt was made to discern methods of data mining (decision trees) and statistical methods (logistic regression). The aim of the comparison is to choose the method that best classifies patients, thus providing a very beaver prediction quality of the constructed classifier. The results compared the calculated measures of the quality of constructed classifiers.

Dataset contains clinical data of 152 patients affected by Ulcerative Colitis (UC) and other colon diseases. Patients are characterized by 117 attributes and classified into two groups: patients with ulcerative colitis and patients with other diseases of the digestive system, which is not a coexisting disease for the disease under examination. Our goal was to find rules which help to reclassify patients from one group to group of not healthy persons.

Too many variables can negatively impact the performance of the model. As a consequence, the first stages of the study, during which initial data processing is performed, are important. The data can be subjected to selection, transformation, or delete unwanted variables.

After removing variables where the percentage of missing data exceeded 60%, the number of attributes decreased. There are 73 attributes left. Subsequently, all the attributes associated with treatment were excluded from the analysis, since predicates describing the treatment cannot determine the occurrence of the disease. Finally, there are 152 cases that describe 64 attributes to the next stages of analysis.

In order to reduce the amount of data, they are subjected to selection methods to obtain the set of attributes that are most strongly associated with the classifier (dependent). Selected attributes are following: *age, smoke, blood_feces, eosinophils, AlAT, sodium, potassium* (Table 3).

Due to the fact that the classification models are very sensitive to data gaps, the k-nearest neighbors have been filled in (missing parameters = k neighbors = 3, number of patterns = 10).

Table 3. Classification attributes

Attribute	Value
Age	Numeric
Smoke	{0-no, 1-yes}
Blood feces	{0-no, 1-yes}
Eosinophils	Numeric
AlAT	Numeric
Sodium	Numeric
Potassium	Numeric

J48 classifier

The J48 method uses the C4.5 algorithm to generate a decision tree. The C4.5 algorithm divides the original dataset from each attribute. The obtained confusion matrix shown in the Table 4.

Table 4. Confusion matrix

Observed effects	Expected effects	
	Ulcerative colitis	No ulcerative colitis
Ulcerative colitis	78	8
No ulcerative colitis	8	58

The first stage of the modeling process was the evaluation of its most important statistics. This allowed for the preliminary determination of the accuracy of the decision rules generated and provided information on the type of possible errors. Table 5 shows a summary of basic statistics defining the quality assessment of the J48 model.

Table 5. Statistics for J48 model

Factor	Value
Correctly classified instances	89%
Incorrectly classified instances	11%
Kappa statistic	0.7858

The percentage of correctly classified attributes by the decision tree is 89%. It is a high, correct result and indicates good quality of the generated decision tree. The Kappa Statistics indicator is relatively low, which means that there are nearly 22% of observations with which the random classifier did not manage (Table 5).

Table 6 presents a set of quality measures for the J48 model. For the classification attribute characterizing the lack of relapse, the value of the true positive meter turned out to be very high (91%), which is a satisfactory result. The false positive rate is lower (12%), which may indicate a sufficient quality of the generated model.

Table 6. Measures for J48 model

Class	TP rate	FP rate	Precision	Recall	F-Measure	AUC
Ulcerative colitis	0.907	0.121	0.907	0.907	0.907	0.969
No ulcerative colitis	0.876	0.093	0.879	0.879	0.879	0.969

The F-Measure measure, which estimates the overall quality of the model, has a high value (91%) and indicates a correctly built medical process.

Random Tree

The Random Tree method of classification of decision trees is an algorithm that combines decision trees and random forest methods (Random Forest). The model of the decision tree itself is sufficient when a smaller range of variables is examined.

The below tables show the confusion matrix (Table 7) and statistics (Table 8) on the quality evaluation of the modeled process. The obtained value of correctly classified attributes is very high (98%), which suggests very good quality of the tree. The

Table 7. Confusion matrix

Observed effects	Expected effects	
	Ulcerative colitis	No ulcerative colitis
Ulcerative colitis	86	0
No ulcerative colitis	2	64

assessment of the goodness of the model is confirmed by the high Kappa statistic. This fact informs about a high, 97% of the number of classified observations.

Table 8. Measures for Random Tree model

Factor	Value
Correctly classified instances	98%
Incorrectly classified instances	2%
Kappa statistic	0.9731

Table 9. Measures for Random Tree model

Class	TP rate	FP rate	Precision	Recall	F-Measure	AUC
Ulcerative colitis	1	0.03	0.977	1	0.989	0.985
No ulcerative colitis	0.97	0	1	0.97	0.985	0.985

Table 9 contains a set of quality measures for the modeled process. The attribute representing no relapse in the true positive meter category has a value of 100%. The false positive meter is an absolutely different value close to 3%. Based on these two values, we can conclude a very good fit of the modeled process. Other indicators (Precision, Recall and F-Measure) also prove the accuracy of the above conclusion.

Logistic regression

After constructing a formal logistic regression model, its structural parameters were estimated. In Table 10 we have presented the results of the estimation.

The model predicted 46% of the variation in terms of the dependent variable, based on Nagelkerke R-squared. Based on the analysis, it was found that the model indeed provides a dependent variable: $\chi 2(5) = 38.82$; $p = 0.01$. Based on the criterion of Hosmer and Lemeshow it found that the model was good suited to the data: $\chi 2$ (8) = 9.25; $p = 0.421$.

Based on the results of the Wald test, it was found that age significantly affected the occurrence of illness W (1) = 6.55; $p = 0.01$ (result on the border of statistical trends). Age highest the chance of ilness based on B = 0.03, but it is very small value.

Variable 'AIAT' also significantly affected the occurrence of ulcerative colitis W (1) = 12.34; $p = 0.0004$. AIAT increase the chance of disease based on B = 0.06.

As we can see the dependent variable is significantly affected by the level of sodium (W (1) = 11.03; $p = 0.0009$). According to the results the variable increases the chance of ilness based on B = 0.32.

The frequency of smoke significantly affects the occurrence of ulcerative colitis W (1) = 142.94; $p < 0.000$. The smoke increases the chance of disease (B = 5.08 for "yes" and B = 5.53 for "no"). As we can see, a smoker is less likely to get sick.

The intensity of number of blood feces also significantly affects the occurrence of disease W (1) = 214.59; $p < 0.000$. The intensity decreases the chance of illness (B = −7.29 and B = −5.73). This means that increased blood in the stool increases the likelihood of getting sick.

Table 10. Parameters of the model – Wald test

	Parameter B	Standard error	Wald statistic	95% CI of Exp (B)		p-value
				High	Low	
Age	0.0338	0.01319	6.5548	0.0079	0.0596	0.010460
AIAT	0.0562	0.01601	12.3409	0.0249	0.0876	0.000443
Sodium	0.3176	0.09562	11.0319	0.1302	0.5050	0.000896
Smoke (1)	5.0761	0.42458	142.9370	4.2440	5.9083	0.000000
Smoke (0)	5.5347					
Blood_feces (0)	−7.2963	0.49808	214.5892	−8.2725	−6.3201	0.000000
Blood_feces (1)	−5.7304					
Constant	0.0338	0.01319	6.5548	0.0079	0.0596	0.010460

Tables 11, 12 and 13 show statistics on the quality evaluation of the modeled process. The obtained value of correctly classified attributes is very high (82%), which suggests very good quality of classification. The attribute representing no relapse in the true positive meter category has a value of 86%. The false positive meter is an absolutely different value close to 24%. Based on these two values, we can conclude a good fit of the modeled process. Other indicators (Precision, Recall and F-Measure) also prove the accuracy of the above conclusion. The ROC curve indicates that the built classifier very well classifies cases into the appropriate group (category 0 - sick). AUC = 0.919 confirms the above statement.

Table 11. Parameters for the model

Observed effects	Expected effects	
	Ulcerative colitis	No ulcerative colitis
Ulcerative colitis	74	12
No ulcerative colitis	16	50

Table 12. Model parameters

Correctly classified instances	124	81.5789%
Incorrectly classified instances	28	18.4211%

Table 13. Model parameters

Class	TP rate	FP rate	Precision	Recall	F-Measure	AUC
Ulcerative colitis	0.86	0.242	0.822	0.86	0.841	0.919
No ulcerative colitis	0.758	0.14	0.806	0.758	0.781	0.919

5 Conclusion

In this paper, we built classification models for dependent variable. It becomes important to refer to the features that will enable you to get the highest impact on patient's recovery. This paper discusses the most commonly used classification algorithms. Classification methods are widely used in medicine. In our work, we attempted to build a classifier that would classify patients undergoing ulcerative colitis and other conditions within the lower gastrointestinal tract. We calculated the basic statistics of each model. Random Tree turned out to be the best (98% correctly classified instances), so in the future be used to build a recommendation system for hospitals.

This work was supported by MB/WM/8/2016 and financed with use of funds for science of MNiSW. The Bioethical Commission gave permission for the analysis and publication of results.

References

1. Bender, R., Grouven, U.: Logistic regression models used in medical research are poorly presented [Letter]. BMJ **313**, 628 (1996)
2. Breiman, L., Friedman, J.H., Olshen, R.A., Stone, C.J.: Classification and Regression Trees. Wadsworth International Group, Belmont (1984)
3. Cabena, P., Hadjinian, P., Stadler, R., Zanassi, A.: Discovering Data Mining: From Concept to Implementation. Prentice Hall, Upper Saddle River (1998)
4. Campillo, C.: Standardizing criteria for logistic regression models. Ann. Intern. Med. **119**, 540–541 (1993)
5. Cheng, J., Greiner, R.: Learning Bayesian belief network classifiers: algorithms and system. In: Stroulia, E., Matwin, S. (eds.) Advances in Artificial Intelligence, AI 2001. LNCS (LNAI), vol. 2056, pp. 141–151. Springer, Heidelberg (2001). https://doi.org/10.1007/3-540-45153-6_14
6. Chin, S.: The rise and fall of logistic regression. Aust. Epidemiol. **8**(3), 7–10 (2001)
7. Dardzinska, A.: Action Rules Mining. SCI. Springer, Heidelberg (2013). https://doi.org/10.1007/978-3-642-35650-6
8. Frawley, W., Piatetsky-Shapiro, G., Matheus, C.: Knowledge discovery in databases: an overview. Knowl. Discov. Databases 1–27 (1991)
9. Hall, G.H., Round, A.P.: Logistic regression: explanation and use. J. Roy. Coll. Phys. Lond. **28**, 242–246 (1994)
10. Hand, D., Mannila, H., Smyth, P.: Eksploracja danych. Wydawnictwa Naukowo – Techniczne, Warszawa, pp. 35–61, 91–127, 181–201 (2005)
11. Harrell, F.: Regression Modeling Strategies with Applications to Linear Models, Logistic Regression, and Survival Analysis. Springer-Verlag, New York (2001)
12. Hosmer, D., Lemeshow, S.: Applied Logistic Regression. Wiley, New Jersey (2000)
13. Jiang, H., Kulkarni, P.M., Mallinckrodt, C.H., Shurzinske, L., Molenberghs, G., Lipkovich, I.: To adjust or not to adjust for baseline when analyzing repeated binary responses? The case of complete data when treatment comparison at study end is of interest. Pharm. Stat. **14**, 262–271 (2015)
14. de Jong, P., Heller, G.Z.: Generalized Linear Models for Insurance Data. Cambridge University Press, Cambridge (2008)

15. Kasperczuk, A., Dardzinska, A.: Comparative evaluation of the different data mining techniques used for the medical database. Acta Mech. Autom. **10**(3), 233–238 (2016)
16. Khan, K.S., Chien, P.F., Dwarakanath, L.S.: Logistic regression models in obstetrics and gynecology literature. Obstet. Gynecol. **93**, 10014–10020 (2000)
17. Koronacki, J., Mielniczuk, J.: Statystyka, p. 206. WNT, Warszawa (2006)
18. Larose, D.T.: Odkrywanie wiedzy z danych. Wprowadzenie do eksploracji danych. Wydawnictwo Naukowe PWN, Warszawa (2013)
19. Levy, P.S., Stolte, K.: Statistical methods in public health and epidemiology: a look at the recent past and projections for the next decade. Stat. Methods Med. Res. **9**, 41–55 (2000)
20. Morzy T.: Eksploracja danych, pp. 10–125, 196–325. Wydawnictwo Naukowe PWN, Warszawa (2013)
21. Quinlan, J.R.: Introduction of decision trees. Mach. Learn. **1**, 81–106 (1986)
22. Zhang, Z., Chen, K., Ni, H., et al.: Predictive value of lactate in unselected critically ill patients: an analysis using fractional polynomials. J. Thorac. Dis. **6**, 995–1003 (2014)

Application of Regularized Online Sequential Learning for Glucose Correction

Hieu Trung Huynh[1,2(✉)] and Yonggwan Won[3]

[1] Faculty of Engineering, Vietnamese-German University, Ho Chi Minh City,
Vietnam
hthieu@ieee.org
[2] Faculty of Information Technology, Industrial University of Ho Chi Minh City,
Ho Chi Minh City, Vietnam
[3] Department of Computer Engineering, Chonnam National University,
Gwangju 500-757, Korea
ykwon@chonnam.ac.kr

Abstract. Glucose measurement by using handheld devices is applied widely due to their comfortabilities. They are easy to use and can give results quickly. However, the accuracy of measurement results is affected by interferences, in which hematocrit (HCT) is one of the most highly affecting factors. In this paper, an approach for glucose correction based on the neural network is presented. The regularized online sequential learning is utilized for hematocrit estimation. The transduced current curve which is produced by the chemical reaction during glucose measurement is used as an input feature of neural network. The experimental results shown that the proposed approach is promising.

Keywords: Hematocrit · Neural network · Online training ·
Glucose correction · Handheld device

1 Introduction

Diabetes mellitus is one of the leading diseases worldwide. It relates to several the long-term complications including cardiovascular, hypoglycemia diabetic ketoacidosis, hyperosmolar, retinopathy, neuropathy, and nephropathy. The current treatment methods for insulin dependent diabetes such as continuous infusion of insulin or subcutaneous insulin injection require frequently evaluating the variation of glucose concentration.

The major tools for managing the glucose concentration are the point-of-care (POC) or handheld blood glucose meters. They are easy to use and relatively cheap, however their accuracy is affected by various interferences, in which the hematocrit is one of the most highly affecting factors for POC and handheld glucose measurements [1, 2]. It was reported that a low hematocrit is associated with overestimation, while a high hematocrit is associated with underestimation of glucose results [3–5]. Hence, one of approaches to improve the accuracy of glucose measurement is to reduce the effects of hematocrit level. The hematocrit level can be estimated by employing commercial impedance analyzers with traditional centrifugation measurements or by dielectric spectroscopy [6]. These approaches are in vitro, quite complicated or require individual

© Springer-Verlag GmbH Germany, part of Springer Nature 2019
A. Hameurlain et al. (Eds): TLDKS XLI, LNCS 11390, pp. 160–171, 2019.
https://doi.org/10.1007/978-3-662-58808-6_7

devices, they cannot be applied in the handheld devices. In this study, we present an approach for hematocrit estimation based on neural networks which are trained by the regularized online sequential algorithm.

The neural network is widely applied in several applications [7–10] due to its abilities to solve problems which are difficult to handle by using traditional approaches and to approximate complex nonlinear mappings directly from input patterns. Several network architectures have been developed, however it was shown that the single hidden layer feedforward neural networks (SLFN) can approximate any function if the activation function is chosen properly. Hence, in this study, we have investigated in the SLFN for biomedical processing. Several training algorithms have been developed for SLFNs, in which one of the effective ones is extreme learning machine (ELM) [11, 12]. This algorithm can obtain good performance with higher learning speed in many applications. Besides batch learning types, sequential learning algorithms are preferred for neural networks in many applications, they do not require the fully available training set and do not require retraining whenever a new training data received. In this paper, the neural network is trained by the regularized online sequential learning algorithm. The rest of this paper is organized as follow. Section 2 presents the materials and methods. The experimental results and analysis are shown in Sect. 3. Finally, we make the conclusion in Sect. 4.

2 Materials and Methods

2.1 Transduced Current Curves

In this study, we focus on the handheld glucose meters using the biosensors. These biosensors use enzymes to break the blood glucose down. One of enzymes commonly used to detect the glucose levels is the glucose oxidase (GOD), it catalyzes the oxidation of glucose by oxygen to produce gluconic acid and hydrogen peroxide.

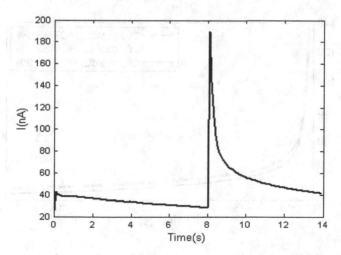

Fig. 1. Anodic current curve

$$\text{Glucose} + O_2 + \text{GO/FA} \rightarrow \text{Gluconic acid} + H_2O_2 + \text{GO/FADH}_2$$

$$\text{GO/FADH}_2 + \text{Ferricinium}^+ \rightarrow \text{GO/FAD} + \text{Ferricinium}$$

$$\text{Ferrocence} \rightarrow \text{Ferrocence}^+ + e^-.$$

The reduced form of the enzyme (GO/FADH$_2$) is oxidized to its original state by an electron mediator (ferrocence). The active electrode then oxidizes the resulting reduced mediator to produce the transduced anodic current. The transduced anodic current curve obtained in the first 14 s is represented in Fig. 1 [13]. It was shown that the first eight seconds do not contain the information of hematocrit and glucose level; it may be an incubation time for waiting the enzyme reaction to be activated. In our study, we concentrate on the second part of the current curve during the next six seconds. In the period of the next six seconds, the anodic current curve is sampled at a frequency of 10 Hz to produce current points.

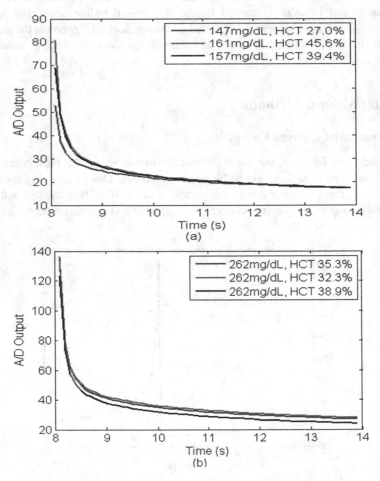

Fig. 2. Effects of hematocrit on glucose measurement: (a) same measured value on current curve but different glucose values, (b) different measured values on current curves but same glucose value.

The glucose values can be obtained from this transduced current curve. However, it was shown that the accuracy is affected by the hematocrit is the most highly effecting factor. Figure 2 illustrates a case where three current curves from time point 11.5 s to 14 s as shown in Fig. 2a. It provides the same value (the measured values of three curves at time point 14 s are the same as 17.3439) even though glucose values corresponding to these current curves are different those are 147 mg/dL, 161 mg/dL and 157 mg/dL for the hematocrit of 27%, 45.6% and 39.4%, respectively. In Fig. 2b, the same glucose value of 262 mg/dL, the measured values on three current curves corresponding to different hematocrit levels are different.

Other researches also demonstrated the relationship between the errors of glucose measurement and hematocrit levels [6, 13, 14]. The result from Louie et al. [15] also shows that the accuracy of glucose measurement can be improved if the effects of hematocrit is reduced. However, we cannot apply the traditional methods for hematocrit estimation. In this study, an approach using the transduced current curves is proposed.

2.2 Neural Networks Trained by Online Training Algorithms for Hematocrit Estimation

The vector of d current points sampled from the second part of the j-th current curve can be denoted as $\mathbf{x}_j = [x_{j1}, x_{j2}, \ldots, x_{jd}]$. This vector is used as the input features of the neural network for estimating hematocrit. The architecture neural network used in this study is single hidden layer feedforward neural network (SLFN) which can approximate any function if the number of hidden nodes and the activation function are chosen properly. The typical architecture of SLFN is shown in Fig. 3, which includes d input nodes, N hidden nodes and C output nodes.

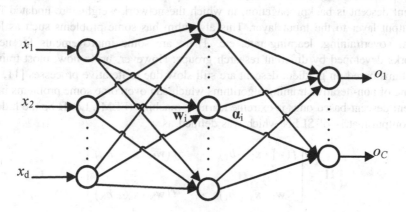

Fig. 3. The architecture of SLFN

Let $f(\cdot)$ be the activation function of hidden units. Mathematically, the SLFNs can be modeled as:

$$\mathbf{o} = \sum_{i=1}^{N} \alpha_i f(\mathbf{w}_i \cdot \mathbf{x} + b_i), \ \mathbf{x} \in \mathbb{R}^d, \tag{1}$$

where $\mathbf{o}$ is the output vector, $\mathbf{w}_i = [w_{i1}, w_{i2}, \ldots, w_{iN}]$ is the input weight vector connecting from the input units to the i-th hidden unit, α_i is the weight vector connecting from the i-th hidden unit to the output units, and b_i is the threshold of the i-th hidden unit, $\mathbf{w}_i \cdot \mathbf{x} = \ <\mathbf{w}_i, \mathbf{x}>\ $ is the inner product of $\mathbf{w}_i$ and $\mathbf{x}$. One of big problems in neural networks is training.

Given n training patterns (x_j, t_j), $j = 1, 2, \ldots, n$, where $xj = [x_{j1} x_{j2} \ldots x_{jd}]T$ and $t_j = [t_{j1} t_{j2} \ldots t_{jC}]T$ are the j-th input pattern and its target, respectively. The main goal of training process is to determine the network weights $\mathbf{w}_i$, α_i, and biases b_i that minimize the error function defined by

$$E = \sum_{j=1}^{n} (\mathbf{o}_j - \mathbf{t}_j)^2, \tag{2}$$

where o_j is the output vector corresponding to the j-th input pattern. Traditionally, this task is performed based on the gradient descent, in which the network weights $\mathbf{g}$ (consisting of $\mathbf{w}$, α and b) are updated iteratively by:

$$\mathbf{g}_k = \mathbf{g}_{k-1} - \eta \frac{\partial E}{\partial \mathbf{g}}, \tag{3}$$

where η is the learning rate. One of the most popular training algorithms based on gradient descent is backpropagation, in which the network weights are updated from the output layer to the input layer. This algorithm has some problems such as local minima, overtraining, learning rate, etc. There are some improvements for neural networks developed by different research groups. However, up to now, most training algorithms based on gradient descent are still slow due to iterative processes [11, 12].

One of non-iterative training algorithms which can overcome some problems in the gradient descent-based ones is extreme learning machine (ELM). Let $\mathbf{H}$ be the hidden-layer-output matrix of SLFN which was defined as

$$\mathbf{H} = \begin{bmatrix} f(\mathbf{w}_1 \cdot \mathbf{x}_1 + b_1) & \cdots & f(\mathbf{w}_N \cdot \mathbf{x}_1 + b_N) \\ \vdots & \ddots & \vdots \\ f(\mathbf{w}_1 \cdot \mathbf{x}_n + b_1) & \cdots & f(\mathbf{w}_N \cdot \mathbf{x}_n + b_N) \end{bmatrix}, \tag{4}$$

The main goal in ELM is to determine the network weights based on the linear model defined by

$$\mathbf{HA} = \mathbf{T}, \tag{5}$$

where $\mathbf{T} = [\mathbf{t}_1 \mathbf{t}_2 \ldots \mathbf{t}_n]^T$, $\mathbf{A} = [\boldsymbol{\alpha}_1 \boldsymbol{\alpha}_2 \ldots \boldsymbol{\alpha}_N]^T$. In the ELM, the input weights and biases of hidden units are randomly assigned, and the output weights are determined by

$$\hat{\mathbf{A}} = \mathbf{H}^\dagger \mathbf{T}, \tag{6}$$

where $\mathbf{H}^\dagger$ is the pseudo-inverse of $\mathbf{H}$.

When the training data is very large or not available fully, the online training approaches should be addressed. An online training method based on the ELM called sequential extreme learning machine (OS-ELM) was proposed by Liang *et al.* [16]. The OS-ELM supposes that $\mathbf{H}^T\mathbf{H}$ is nonsingular and pseudo-inverse of $\mathbf{H}$ is given by

$$\mathbf{H}^\dagger = \left(\mathbf{H}^T\mathbf{H}\right)^{-1}\mathbf{H}^T. \tag{7}$$

From above assumptions, the output weights are updated by following rules:

$$\mathbf{A}_k = \mathbf{A}_{k-1} + \mathbf{L}_k^{-1}\mathbf{H}_k^T(\mathbf{T}_k - \mathbf{H}_k\mathbf{A}_{k-1}), \tag{8}$$

$$\mathbf{L}_k = \mathbf{L}_{k-1} + \mathbf{H}_k^T\mathbf{H}_k \tag{9}$$

where $\mathbf{T}_k = [\mathbf{t}_{\sum_{i=0}^{k-1} n_i + 1} \ \mathbf{t}_{\sum_{i=0}^{k-1} n_i + 2} \ldots \mathbf{t}_{\sum_{i=0}^{k} n_i}]^T$, $\mathbf{H}_k = [\mathbf{h}_{\sum_{i=0}^{k-1} n_i + 1} \ \mathbf{h}_{\sum_{i=0}^{k-1} n_i + 2} \cdots \mathbf{h}_{\sum_{i=0}^{k} n_i}]^T$, and $\mathbf{h}_j = [f(\mathbf{w}_1 \cdot \mathbf{x}_j), \ldots, f(\mathbf{w}_N \cdot \mathbf{x}_j)]^T$. The initialization of $\mathbf{A}_k$ corresponding to an initial training set $\mathbf{S}_0 = \{(\mathbf{x}_j, \mathbf{t}_j) | j = 1, \ldots, n_0\}$ is given by

$$\mathbf{A}_0 = \mathbf{L}_0^{-1}\mathbf{H}_0^T\mathbf{T}_0, \tag{10}$$

where $\mathbf{L}_0 = \mathbf{H}_0^T\mathbf{H}_0$, $\mathbf{T}_0 = [\mathbf{t}_1\mathbf{t}_2 \ldots \mathbf{t}_{n0}]^T$, and $\mathbf{H}_0 = [\mathbf{h}_1\mathbf{h}_2 \ldots \mathbf{h}_{n0}]^T$. In summation, the OS-ELM algorithm is described as follows:

(1) **Initialization:**
   For the initial training subset $\mathbf{S}_0 = \{(\mathbf{x}_j, \mathbf{t}_j) | j = 1, \ldots, n_0\}$,

   – Assign random values for $\mathbf{w}$'s and b's.
   – Calculate hidden layer output matrix $\mathbf{H}_0$.
   – Determine $\mathbf{L}_0$ and then $\mathbf{A}_0$ using by using Eq. 10.

(2) **Updating weight:** For the arriving training subset
   $\mathbf{S}_k = \{(\mathbf{x}_j, \mathbf{t}_j) | j = \sum_{i=0}^{k-1} n_i + 1, \ldots, \sum_{i=0}^{k} n_i\}$,

   – Determine $\mathbf{H}_k$.
   – Determine $\mathbf{L}_k$ by Eq. 9.
   – Update the output weights $\mathbf{A}_k$ by Eq. 8.

In the first step of algorithm (*initialization*) the input weights and biases are assigned by random values; then the output weight matrix A_0 is computed. Following the initialization step, the updating process is performed, in which the output weights are updated for each arriving data of one-by-one or chunk-by-chunk.

In the real applications, the collected data are often included noise. Hence, the risk minimization as shown in (2) may lead to a poor generalization. One of approaches which can overcome this problem is to optimize the norm of output weight vector. The solution for A of Eq. 5 can be replaced by seeking A that minimizes

$$\|HA - T\|^2 + \lambda\|A\|^2, \tag{11}$$

where $\|\cdot\|$ is Euclidean norm and λ is a positive constant. The solution for A from Eq. 11 is given by

$$\hat{A} = (H^T H + \lambda I)^{-1} H^T T. \tag{12}$$

The learning rules for online sequential learning process were given by Huynh et al. [17]. For an initial training set $S_0 = \{(x_j, t_j) | j = 1, \ldots, n_0\}$, the output weights are initialized by

$$A_0 = L_0^{-1} H_0^T T_0, \tag{13}$$

where $L_0 = H_0^T H_0 + \lambda I$, $T_0 = [t_1 t_2 \ldots t_{n0}]^T$, and $H_0 = [h_1 h_2 \ldots h_{n0}]^T$. In the updating phase, the output weights are updated by

$$U_k = U_{k-1} - U_{k-1} H_k^T (I + H_k U_{k-1} H_k^T)^{-1} H_k U_{k-1} \tag{14}$$

$$A_k = A_{k-1} + U_k H_k^T (T_k - H_k A_{k-1}). \tag{15}$$

where

$$U_0 = (H_0^T H_0 + \lambda I)^{-1}$$
$$= \frac{1}{\lambda} I - \frac{1}{\lambda} H_0^T (\lambda I + H_0 H_0^T)^{-1} H_0 \tag{16}$$

2.3   Glucose Correction

Let g_p, g_r be the measured glucose values from portable device and reference machine (YSI2700 or YSI2300), respectively. The residual r_p is given by

$$r_p = g_p - g_r. \tag{17}$$

The main goal of correction process is to find mapping

$$f : g_p \rightarrow g_c,$$ (18)

where g_c is the corrected glucose values corresponding to g_p. The proposed mapping for f is given by:

$$g_c = g_p - r_p.$$ (19)

It is important to find a function g mapping from hematocrit to residual r_p as follows g: $HCT_p \rightarrow r_p$

$$r_p = g(HCT_p),$$ (20)

where HCT_p is hematocrit estimated from the transduced current curve. The simplest approach is to apply a linear model. The mapping function g is represented by:

$$r_p^j = g(HCT_p^j) = a_g \text{x} HCT_p^j + b_g,$$ (21)

where a_g and b_g are two parameters which must be determined. For n training samples $(HCT_p^j, r_p^j), j = 1, 2, \ldots, n$. The parameters can be approximated with minimum error of the following equation:

$$\mathbf{r} = \mathbf{H}_{\text{HCT}} \mathbf{a}_g,$$ (22)

where $\mathbf{r} = [r_p^1 \, r_p^2 \, \cdots \, r_p^n]^T$, $\mathbf{a}_g = [a_g \, b_g]^T$, and $\mathbf{H}_{\text{HCT}} = \begin{bmatrix} HCT_p^1 & HCT_p^2 & \cdots & HCT_p^n \\ 1 & 1 & \cdots & 1 \end{bmatrix}^T$.

The least mean square error solution for (22) is given by

$$\hat{\mathbf{a}}_g = (\mathbf{H}_{\text{HCT}}^T \mathbf{H}_{\text{HCT}})^{-1} \mathbf{H}_{\text{HCT}}^T \mathbf{r}.$$ (23)

The residual for any sample is given by

$$r_p = [HCT_p 1] \hat{\mathbf{a}}_g.$$ (24)

3   Experimental Results

In this study, we evaluate the performance on the dataset which was obtained from 191 blood samples. These samples were obtained from randomly selected volunteers. There are four measurements for each sample including (1) the accurate hematocrit using centrifugation method, (2) accurate glucose using YSI2700, (3) glucose values using a handheld device, and (4) the anodic current curves. From the second part of curve, which is after the incubation time, fifty-nine current points are sampled at a frequency of 10 Hz. The dataset was divided into two subsets, in which the forty percent of dataset is used for training and the sixty percent is used for blind testing. In our experiment, the neural networks were trained by the OS-ELM, our proposed method and offline ELM. The number of hidden units was 12 for ELM and online training algorithms.

3.1   Hematocrit Estimation

The hematocrit values collected from centrifugation method have the distribution as shown in Fig. 4, in which the mean is 36.02 and the deviation is 6.39. The average result of fifty trials with the whole current curve is shown in Table 1. The root mean square error (RMSE) is computed by

$$RMSE = \sqrt{\frac{1}{n}\sum_{j=1}^{n}\left(o_j - t_j\right)^2} \tag{25}$$

where o_j is the estimated value and t_j is the reference value.

Table 1. Comparison with reference hematocrit measurements using centrifugation

Method	Training		Testing		# nodes
	RMSE	Std	RMSE	Std	
ELM (offline)	3.67	0.34	4.49	0.51	12
OS-ELM [16]	3.69	0.26	4.37	0.37	12
Proposed approach	3.65	0.28	4.18	0.35	12

From the Table 1 we can see that the accuracy of the proposed method corresponding to the testing set is 4.18 which is compatible to that of the offline training

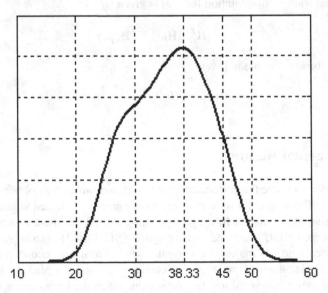

Fig. 4. Distribution of collected hematocrit

methods for the same number of hidden nodes. Note that, for the online training method, the devices can be still trained with new samples during the using process which can expect to improve the performance further.

3.2   Glucose Correction

There is a relationship between hematocrit and residuals which are defined as differences of handheld glucose measurements minus the YSI2007 glucose measurements. In addition, using the test statistic for the slope given by

$$t_{slope} = \frac{slope}{\sigma_{slope}} \tag{26}$$

and using the P-test we see that the slope value is significantly different than 0 (p < 0.01). Therefore, we can conclude that effect of hematocrit on handheld measurements is significant which is consistent with the previous reports. The RMSE for handheld on the test set without error correction is 16.4149 while that with error correction is 13.53. The t-test for slope without error correction is −3.846 (p-value < 0.001) which shows dependence of residuals on hematocrit levels, while the t-test for slope with the error correction is 0.23, these results show that the effects of hematocrit are reduced after error correction.

On the error tolerance of ±15 mg/dL for glucose levels ≤ 100 mg/dL and p% for glucose levels >100 mg/dL, Table 2 presents comparison results (within the error tolerance) of error correction corresponding different values of p.

Table 2. Comparison results on different criteria of error tolerance.

p(%)	Before error correction	After error correction
15	91.60%	94.66%
16	94.66%	95.42%
17	94.66%	96.18%
18	95.42%	96.95%
19	96.18%	96.95%
20	96.95%	96.95%

The criteria proposed by the National Committee for Clinical Laboratory Standards are that error tolerances of ±15 mg/dL for glucose levels ≤ 100 mg/dL and ±20% for glucose levels >100 mg/dL. At least 95% of glucose meter measurements should fall within these error tolerances. We can see that both approaches, before and after error correction, have 96.95% of glucose measurements within the error tolerances which satisfies the criteria proposed by the National Committee for Clinical Laboratory Standards. However, the error correction provides an improved performance at levels of error tolerance from 15% to 19%.

4 Conclusion

In this study, we developed an approach for glucose correction in handheld devices by reducing the effect of hematocrit. The hematocrit estimation is performed by using the online sequential method with input features from transduced current curves. The transduced current changing curves are produced by chemical reactions of glucose oxidase in the electrochemical biosensors. The experimental results showed that the online training method is compatible to the offline training methods but note that the accuracy of devices can be still improve during the using process. The accuracy of glucose measurement using electrochemical biosensors is improved after reducing the effects of hematocrit.

References

1. Aynsley-Green, A.: Glucose, a fuel for thought. J. Pediatr. Child Health **27**(1), 21–30 (1991)
2. Hussain, K., Sharieff, N.: The inaccuracy of venous and capillary blood glucose measurement using reagent strips in the new born period and the effect of hematocrit. Early Human Dev. **57**(2), 111–121 (2000)
3. Tang, Z., Lee, J.H., Louie, R.F., Kost, G.J., Sutton, D.V.: Effects of different hematocrit levels on glucose measurements with handheld meters for point of care testing. Arch. Pathol. Lab. Med. **124**(8), 1135–1140 (2000)
4. Kilpatrick, E.S., Rumley, A.G., Myin, H.: The effect of variations in hematocrit, mean cell volume and red blood count on reagent strip tests for glucose. Ann. Clin. Biochem. **30**(5), 485–487 (1993)
5. Kaplan, M., Blondheim, O., Alon, I.: Screening for hypoglycemia with plasma in neonatal blood of high hematocrit value. Crit. Care Med. **17**(3), 279–282 (1989)
6. Treo, E.F., Felice, C.J., Tirado, M.C., Valentinuzzi, M.E., Cervantes, D.O.: Hematocrit measurement by dielectric spectroscopy. IEEE Trans. Biomed. Eng. **25**(1), 124–127 (2005)
7. Lisboa, P.J.G., Ifeachor, E.C., Szczepaniak, P.S.: Artificial Neural Networks in Biomedicine. Springer, Heidelberg (2000). https://doi.org/10.1007/978-1-4471-0487-2
8. Naguib, R.N.G., Sherbet, G.V.: Artificial Neural Networks in Cancer Diagnosis, Prognosis, and Patient Management. CRC Press, Washington D.C (2001)
9. Huynh, H.T., Kim, J., Won, Y.: Performance comparison of SLFN training algorithms for DNA microarray classification. In: Arabnia, H., Tran, Q.N. (eds.) Software Tools and Algorithms for Biological Systems: Advances in Experimental Medicine and Biology, vol. 696, pp. 135–143. Springer, New York (2011). https://doi.org/10.1007/978-1-4419-7046-6_14
10. Huynh, H.T., Kim, J., Won, Y.: Classification study on DNA microarray with feedforward neural network trained by singular value decomposition. Int. J. Biosci. Biotechnol. **1**(1), 17–24 (2009)
11. Huang, G.-B., Zhu, Q.-Y., Siew, C.-K.: Extreme learning machine: a new learning scheme for feedforward neural networks. In: Proceedings of International Joint Conference on Neural Networks. IEEE, Hungary, July 2004
12. Huang, G.-B., Zhu, Q.-Y., Siew, C.-K.: Extreme learning machine: theory and applications. Neurocomputing **70**(1–3), 489–501 (2006)

13. Huynh, H.T., Won, Y., Kim, J.: Neural networks for the estimation of hematocrit from transduced current curves. In: The 2008 IEEE International Conference on Networking, Sensing and Control, pp. 1517–1520. IEEE, Korea (2008)

14. Huynh, H.T., Quan, H.D., Won, Y.: Accuracy improvement for glucose measurement in handheld devices by using neural networks. In: Dang, T.K., Wagner, R., Küng, J., Thoai, N., Takizawa, M., Neuhold, Erich J. (eds.) FDSE 2017. LNCS, vol. 10646, pp. 299–308. Springer, Cham (2017). https://doi.org/10.1007/978-3-319-70004-5_21

15. Louie, R.F., Tang, Z., Sutton, D.V., Lee, J.H., Kost, G.J.: Point of care glucose testing: effects of critical variables, influence of reference instruments, and a modular glucose meter design. Arch. Pathol. Lab. Med. **124**(2), 257–266 (2000)

16. Liang, N.-Y., Huang, G.-H., Saratchandran, P., Sundararajan, N.: A fast and accurate online sequential learning algorithm for feedforward networks. IEEE Trans. Neural Netw. **17**(6), 1411–1423 (2006)

17. Huynh, H.T., Won, Y.: Regularized online sequential learning algorithm for single-hidden layer feedforward neural networks. Pattern Recogn. Lett. **32**(14), 1930–1935 (2011)

Author Index